AF357834

# Temperature Field, Electromagnetic Field, and Operation Control of Permanent Magnet Motor for Electric Vehicles

# Temperature Field, Electromagnetic Field, and Operation Control of Permanent Magnet Motor for Electric Vehicles

Guest Editor

**Zhongxian Chen**

Basel • Beijing • Wuhan • Barcelona • Belgrade • Novi Sad • Cluj • Manchester

*Guest Editor*
Zhongxian Chen
Huanghuai University
Zhumadian
China

*Editorial Office*
MDPI AG
Grosspeteranlage 5
4052 Basel, Switzerland

This is a reprint of the Special Issue, published open access by the journal *World Electric Vehicle Journal* (ISSN 2032-6653), freely accessible at: https://www.mdpi.com/journal/wevj/special_issues/68L9B8CVE9.

For citation purposes, cite each article independently as indicated on the article page online and as indicated below:

Lastname, A.A.; Lastname, B.B. Article Title. *Journal Name* **Year**, *Volume Number*, Page Range.

ISBN 978-3-7258-2945-3 (Hbk)
ISBN 978-3-7258-2946-0 (PDF)
https://doi.org/10.3390/books978-3-7258-2946-0

# Contents

*World Electric Vehicle Journal*

*Editorial*

# Exploring the Drive Motor of Electric Vehicles: Structure, Temperature Rises, and Operational Control of Permanent Magnet Motors

**Zhongxian Chen * and Jiayin Liu**

School of Intelligence Manufacturing, Huanghuai University, Zhumadian 463000, China; liujiayin@huanghuai.edu.cn

* Correspondence: chenzhongxian@huanghuai.edu.cn

**Abstract:** The drive motor is a critical component of electric vehicles. The design of its structure, along with the regulation of temperature rises and operational control, significantly impact the overall performance of electric vehicles, affecting aspects such as dynamic response, anti-interference capabilities, speed stability, and efficiency. This paper provides a review and comprehensive analysis of the Special Issue titled "Temperature Field, Electromagnetic Field, and Operation Control of Permanent Magnet Motor for Electric Vehicles", and discusses the future development directions of permanent magnet motors in electric vehicles. The analysis reveals that the structural design is crucial for both the static and dynamic performance of permanent magnet motors. Additionally, it is imperative to identify an optimal trade-off among various performance parameters in these motors. The paper also verifies several future development directions for permanent magnet motors, including high-speed control, the use of advanced silicon steel sheets, power quality, battery charging efficiency, and energy feedback.

**Keywords:** structure design; temperature rise; control algorithm; permanent magnet motor; electric vehicle

**Citation:** Chen, Z.; Liu, J. Exploring the Drive Motor of Electric Vehicles: Structure, Temperature Rises, and Operational Control of Permanent Magnet Motors. *World Electr. Veh. J.* **2024**, *15*, 483. https://doi.org/10.3390/wevj15110483

Received: 12 October 2024
Accepted: 23 October 2024
Published: 25 October 2024



## 1. Introduction

The advent of permanent magnet materials in the previous century has significantly enhanced the operational performance of motors due to their amalgamation of a high core flux density, high coercivity, and a substantial magnetic energy product [1–3]. These materials are typically integrated into either the rotor or the stator of motors, leading to the design of rotor-type or stator-type permanent magnet motors, respectively. The specific configurations of these motors vary according to the shape and positioning of the permanent magnet materials. Over the last decade, research has demonstrated that the fundamental structure of permanent magnet motors significantly impacts their thermal behavior and electromagnetic properties, despite the advent of various control technologies [4,5]. This paper aims to provide a detailed review and comprehensive analysis of the Special Issue titled "Temperature Field, Electromagnetic Field, and Operation Control of Permanent Magnet Motor for Electric Vehicles", which focuses on cutting-edge technologies and theories pertinent to the use of permanent magnet motors in electric vehicles.

Electric vehicles, characterized by multimodal operation demands such as starting, accelerating, and climbing, require permanent magnet motors that exhibit the following performance characteristics [6–9]:

- Compact size and a light weight;
- High efficiency throughout the whole speed spectrum;
- Low speed and high torque characteristics;
- Constant power within a wide speed range;
- High safety and reliability;

- High voltage;
- High speed;
- High mechanical stress;
- Low temperature rises;
- High power density.

These performance traits underscore the importance of structural design and operational control technology as fundamental elements in the research on the use of permanent magnet motors for electric vehicles, as illustrated in Figure 1. From Figure 1, it is evident that the structural design is not only crucial for optimizing the magnetic flux amplitude, cogging torque, leakage flux, moment of inertia, and $d$-$q$ axis inductance, but also influences the dynamic performance of the motor. Concurrently, operational control technology governs the dynamic response, anti-interference capability, speed stability, and system efficiency of the motor. The power density (kW/kg) of the motor is a function of both the structural design and operational control technology, with a high power density facilitating cost reductions and efficiency enhancements in practical applications.

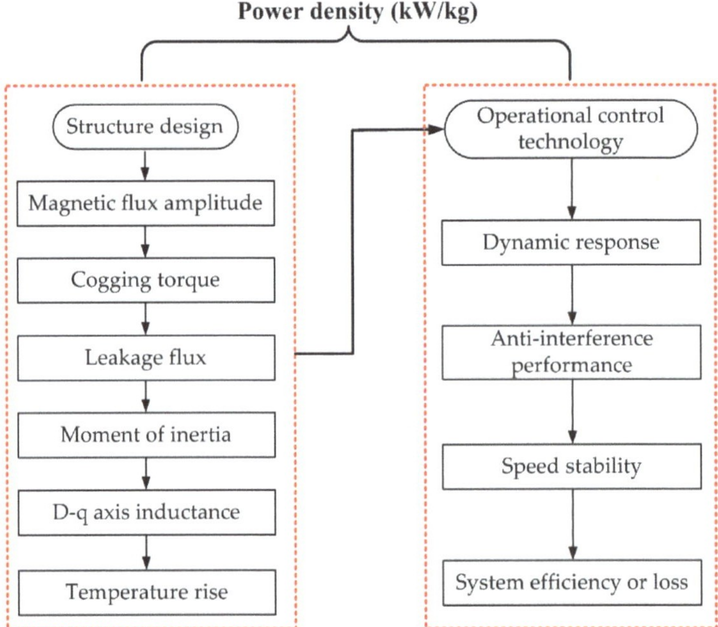


**Figure 1.** The overall process of research on the use of permanent magnet motors for electric vehicles.

Despite many scholars and experts focusing on structural design and operational control technology in the context of permanent magnet motors for electric vehicles, the issue of temperature rises has been largely overlooked. For instance, a review of publications across all MDPI (Multidisciplinary Digital Publishing Institute) journals over the past decade (2014–2024) reveals a significant disparity in research focus:

- Structural design: (27);
- Operational control technology: (14);
- Temperature rises: (1).

Given this research landscape, the objective of this Special Issue is to encourage scholars and experts to intensify their investigations into the drive motors of electric vehicles. This Special Issue welcomes submissions that explore structural design within the constraints of electric vehicles, electromagnetic modeling, computational methods, operational control technologies, and temperature rises. Additionally, contributions on new and emerging topics relevant to the drive motors of electric vehicles are encouraged. While theoretical analyses form the core of the submissions, there is a particular emphasis on practice-oriented research that builds on theoretical foundations.

Fifteen manuscripts were submitted for review in this Special Issue, with ten manuscripts accepted for publication. The principal topics addressed in each paper are summarized in Table 1.

**Table 1.** The main topics addressed in each paper of this Special Issue.

| No | Papers' Authors and Name | Topics Addressed in the Paper |
|---|---|---|
| 1 | Zhang, G.; Jewell, G.W. Combined Electromagnetic and Mechanical Design Optimization of Interior Permanent Magnet Rotors for Electric Vehicle Drivetrains. | Structural design |
| 2 | Wang, H.; Zhang, C.; Guo, L.; Chen, W.; Zhang, Z. Temperature Field Calculation of the Hybrid Heat Pipe Cooled Permanent Magnet Synchronous Motor for Electric Vehicles Based on Equivalent Thermal Network Method. | Temperature rises |
| 3 | Ren, K.; Chen, H.; Sun, H.; Wang, Q.; Sun, Q.; Jin, B. Design and Analysis of a Permanent Magnet Brushless DC Motor in an Automotive Cooling System. | |
| 4 | Li, W.; Xiong, S.; Shi, W. A Novel Battery Temperature-Locking Method Based on Self-Heating Implemented with an Original Driving Circuit While Electric Vehicle Driving: A Numerical Investigation. | |
| 5 | Huang, A.; Chen, Z.; Wang, J. Research on the dq-Axis Current Reaction Time of an Interior Permanent Magnet Synchronous Motor for Electric Vehicle. | Operational control |
| 6 | Geng, Q.; Qin, Z.; Jin, X.; Zhang, G.; Zhou, Z. Direct Torque Control of Dual Three-Phase Permanent Magnet Synchronous Motors Based on Master–Slave Virtual Vectors. | |
| 7 | Chen, Z.; Dai, X.; Faizan, M. Speed Stability and Anti-Disturbance Performance Improvement of an Interior Permanent Magnet Synchronous Motor for Electric Vehicles. | |
| 8 | Geng, Q.; Du, W.; Jin, X.; Zhang, G.; Zhou, Z. Online Fault Detection of Open-Circuit Faults in a DTP-PMSM Using Double DQ Current Prediction. | |
| 9 | Xing, J.; Zhang, J.; Zhuang, X.; Xu, Y. Online Inductance Identification of Permanent Magnet Synchronous Motors Independent of Rotor Position Information. | |
| 10 | Zhao, J.; Zhao, L.; Zou, Y.; Chen, T. Comparison and Analysis of Electromagnetic Characteristics of Basic Structure of Wireless Power Coil for Permanent Magnet Motors in Electric Vehicles. | Emerging topics in the drive system of electric vehicles |

## 2. Structural Design of Permanent Magnet Motors

Permanent magnet motors used in electric vehicles are characterized by their high-speed operation, ranging from a base speed of 4000 RPM to an extended speed of 12,000 RPM. At high speeds, centrifugal forces become a critical constraint, particularly in the rotor section of permanent magnet motors, which lacks additional mechanical containment. Consequently, the structural design phase of the permanent magnet motor necessitates the simultaneous consideration of both centrifugal forces and electromagnetic characteristics.

Paper 1 utilizes a linear elastic model within ANSYS software to analyze the centrifugal forces in the rotor section of a permanent magnet motor, alongside an examination of the motor's electromagnetic characteristics. This model facilitated the optimal structural design of the rotor section, resulting in a final rotor diameter of 165 mm and a shaft length of 103 mm. Moreover, the optimized design concerning electromagnetic characteristics and centrifugal forces enabled the motor to achieve a high power density of 2.4 kW/kg.

During the design of rotor sections with varying diameters, the simulation results indicated that increased power and torque densities in the permanent magnet motor corresponded to an increased leakage flux between the adjacent regions of individual magnet poles. Leakage fluxes are indicative of reduced efficiency in the permanent magnet motor. Consequently, an optimum rotor diameter of 165 mm was determined through a comparative analysis of speed, centrifugal forces, and current density.

Paper 1 also explores the impact of the rotor diameter and shaft length on the hub diameter. For a hub-type permanent magnet motor with a rotor diameter of 165 mm, the calculated electromagnetic torque significantly increases once the hub diameter exceeds 115 mm. This phenomenon is primarily attributed to the magnetic saturation of the leakage flux path in the rotor as the hub diameter increases, thereby enhancing the effective air gap flux and electromagnetic torque between the rotor and stator sections.

Furthermore, an important finding of Paper 1 is that the permanent magnet profile influences the rotor section on both the electromagnetic torque and the maximum localized stress within the motor. Under consistent conditions of speed, dimensions, materials, etc., the simulation results for electromagnetic torque and maximum localized stress with various permanent magnet profiles are detailed in Table 2, and specific rotor section profiles are illustrated in Figure 2. Considering factors such as manufacturing difficulty, cost, and performance requirements, a suitable rotor structure for the permanent magnet motor can be determined using the analytical models and methods.

**Table 2.** The electromagnetic torque and maximum localized stress with different permanent magnet profiles.

| | Permanent Magnet Profile in Figure 2a | Permanent Magnet Profile in Figure 2b | Permanent Magnet Profile in Figure 2c |
|---|---|---|---|
| Electromagnetic torque (Nm) | 239 | 213 | 230 |
| Maximum localized stress (Mpa) | 240 | 229 | 245–261 (with different filler materials) |

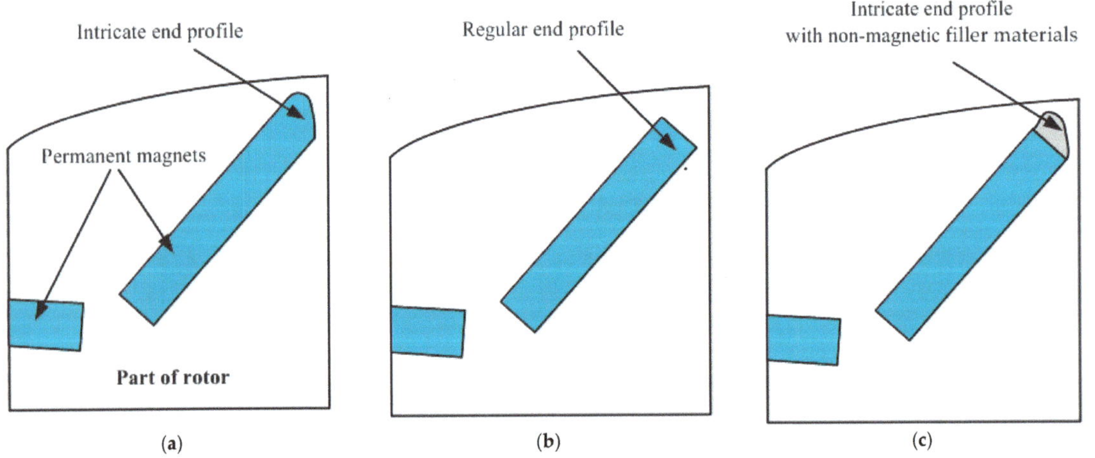


**Figure 2.** The detailed permanent magnet profile of the rotor section of a permanent magnet motor. (**a**) Intricate shape; (**b**) regular shape; (**c**) non-magnetic filler material.

### 3. Temperature Rises in Permanent Magnet Motors

The high torque density and speed inherent in permanent magnet motors lead to a temperature rise in certain internal components. Consequently, developing effective temperature regulation or cooling systems for drive motor systems of electric vehicles has become a crucial research topic.

The conventional method for calculating the temperature rises in permanent magnet motors primarily uses finite element software integrated with computational fluid dynamics (CFD). However, this method has significant drawbacks: it is time-consuming and requires high-performance computing capabilities. Paper 2 introduces a highly efficient method called the equivalent thermal network, which calculates the thermal resistance at each node of the permanent magnet motor. The results validate that the proposed method's calculation error is less than 5%, and it significantly reduces the calculation time to 194 s. Furthermore, Paper 2 offers a detailed analysis of the sources of calculation error, primarily attributing them to the low thermal conductivity of the potting material in the equivalent thermal network model being overlooked.

Paper 3 outlines the design of an automotive cooling system equipped with an adequately structured fan to effectively manage the thermal conditions of permanent magnet motors in electric vehicles. The experimental findings indicate that the internal temperature of the motor increases gradually during the initial operation phase (from 0 to 3300 s) and subsequently stabilizes. Extended operation within the temperature range of 120 °C to 160 °C markedly reduces the motor's operational efficiency. Thus, an active cooling system is indispensable for managing the heat in high-torque and high-speed scenarios in electric vehicle motors.

Moreover, appropriate control technology not only ensures reasonable management of the motor's temperature rise but also helps regulate the battery's temperature to prevent it from exceeding safe limits. Paper 3 proposes a novel space vector pulse width modulation (SVPWM) algorithm with dead zone regulation, termed dSVPWM, to adjust the temperature rises in both the permanent magnet motor and the battery. The simulation results demonstrate that the dSVPWM algorithm effectively maintains the battery's minimum temperature and affirms the feasibility of the motor's speed control. Additionally, compared to the conventional SVPWM algorithm, the dSVPWM algorithm maintains the sinusoidal fluctuations in the phase current and phase difference at consistent levels.

Looking ahead, research on temperature rises in permanent magnet motors should focus on developing a comprehensive analytical model that incorporates both operational and heating efficiencies. This model would facilitate the exploration of temperature-sensitive parameters in permanent magnet motors, thereby enabling the swift and efficient regulation of temperature based on these parameters.

## 4. Operational Control of Permanent Magnet Motor

The primary operating characteristics of permanent magnet motors, such as the speed stability, dynamic response, and anti-disturbance performance, are significantly influenced by the control methods employed. To enhance these characteristics, it is essential to explore and implement two closed-loop control processes: current closed-loop control and speed closed-loop control. Paper 5 introduces an internal model control (IMC) method combined with inverter switching frequency to improve the $dq$-axis current response time. Both the simulation and the experimental results demonstrate that the inverter's switching frequency critically affects the $dq$-axis current response time, as detailed in Table 3. The parameter "Tpwm" represents the inverter switching period. Furthermore, data from Table 3 also indicate that the IMC-PI method offers substantial advantages in terms of the control parameter settings and remains unaffected by the inverter switching period.

**Table 3.** The comparison of $dq$-axis current reaction times when using different control methods.

| Control Methods | Q-Axis Current Reaction Time | |
| --- | --- | --- |
| | **Simulation Results** | **Hardware Experimental Results** |
| PI method | 28 ms (Tpwm = 0.001 s) | 25 ms (Tpwm = 0.001 s) |
| | 135 ms (Tpwm = 0.01 s) | 95 ms (Tpwm = 0.01 s) |
| IMC-PI method | 13 ms (Tpwm = 0.001 s) | 12 ms (Tpwm = 0.001 s) |
| | 13 ms (Tpwm = 0.01 s) | 13 ms (Tpwm = 0.01 s) |

Indeed, the operational performance of permanent magnet motors is determined by the comprehensive application of both current closed-loop and speed closed-loop controls. Papers 6 and 7 adopted the vector control method and direct torque control method, respectively, incorporating these two closed-loop processes to enhance motor performance. In Paper 6, a direct torque control model with a primary–secondary virtual vector was developed and validated, reducing the fluctuations in electromagnetic torque and flux linkages in the permanent magnet motor and significantly decreasing high-order harmonics in the phase current by 69.4%. Paper 7 introduced an extended state observer (ESO) based on sliding mode control (SMC) to improve the speed stability and anti-interference performance of the motor. Compared with the traditional SMC method and linear SMC-ESO method, the experimental results from this paper reveal that only the nonlinear SMC-ESO method effectively enhances both speed stability and anti-interference performance.

In the operational control of permanent magnet motors, dynamic characteristic parameters play a pivotal role in the motors' performance. Consequently, Papers 8 and 9 focus on the online detection or identification of these parameters. Paper 8 proposes a comprehensive detection method that combines open-phase faults (OPF) and open-switching faults (OSF) to enhance the dynamic response capability and anti-disturbance performance of dual three-phase permanent magnet motors. The test results confirm the validity of the proposed method, with an accuracy rate of 97%. Paper 9 utilizes the recursive least squares algorithm and particle swarm optimization to identify the $d$-axis and $q$-axis inductances. The accurate identification of these inductances allows for the stable and precise speed regulation of the motor using the sensorless control method.

## 5. Future Research of Drive Motor for Electric Vehicles

The total drive system of electric vehicles encompasses the inverter, controller, drive motor, sensors, and software. While this Special Issue focuses on the structural design, temperature rises, and operational control of permanent magnet motors, it is imperative to extend research to other related areas, such as power quality, battery charging efficiency, and energy feedback devices [10,11].

For instance, Paper 10 investigates the wireless power transmission performance used for powering the drive system of electric vehicles. Through the structural design of the coil, the study elucidates the mechanism by which the coupling coefficient can enhance the transmission efficiency of wireless power. The simulation and hardware test results demonstrate that, for coils with an equal area or perimeter, the performance of circular coils surpasses that of rectangular coils. Table 4 presents the hardware test results of a comparison of the wireless power transmission performance (coupling coefficient) of circular and rectangular coils. However, a significant finding from Table 4 is that the optimal coupling coefficient for a given type of coil predominantly depends on the coil spacing. Additionally, as all simulation and hardware tests were conducted under the same conditions of wireless power transmission, future research should explore other vital variable characteristics, such as frequency and magnetic interference.

**Table 4.** The wireless power transmission performance of a circular coil and rectangular coil.

| Coil Type | Characteristic | Pacing (Distance Between Transmitter and Receiver Coils) | Coupling Coefficient |
|---|---|---|---|
| Circular | Equal area | 67.82 mm | 11.5% |
| rectangular | | | 9.66% |
| Circular | Equal perimeter | 79.17 mm | 8.17% |
| rectangular | | | 7.8% |
| Circular | Equal outer diameter | 105.17 mm | 4.46% |
| rectangular | | | 4.29% |

## 6. Conclusions

These papers have analyzed a broad spectrum of topics pertaining to the structural design, temperature rises, and operational control of permanent magnet motors in electric vehicles. The findings underscore the pivotal role of structural design technology in advancing research on various aspects of permanent magnet motors, including temperature regulation, operational control, and energy feedback devices. Despite the significant theoretical analyses and related technical applications explored in this Special Issue, numerous challenges remain to be addressed to meet the growing demands for safety, stability, and dynamic response in electric vehicles.

The focus of the papers in this Special Issue was predominantly on the normal operating speed conditions of permanent magnet motors. Therefore, a primary technical challenge that requires further investigation is the operational performance of these motors at higher speeds exceeding 14,000 rpm. At such high speeds, factors such as the controller's processing speed, rotor position detection, inverter switching frequency, and control algorithms play crucial roles in influencing the electromagnetic performance and temperature regulation of the motor. Additionally, the introduction and application of advanced silicon steel sheets in the rotor and stator cores have facilitated their lightweight design, which is emerging as a key development direction for permanent magnet motors. Moreover, it is imperative to conduct comprehensive research and analyses to identify an optimal balance among cost investment, temperature rises, mechanical stress, power density, speed stability, and anti-disturbance performance in permanent magnet motors.

**Author Contributions:** Conceptualization, Z.C.; material organization, J.L.; writing—original draft preparation, Z.C.; writing—review and editing, J.L. All authors have read and agreed to the published version of the manuscript.

**Funding:** This research work is supported by the Major Science and Technology Special Project of Zhumadian (Grant No. ZMDSZDZX2023001).

**Acknowledgments:** We would like to express our gratitude to all authors who contributed to this Special Issue.

**Conflicts of Interest:** The authors declare no conflicts of interest.

## References

1. Wang, A.; Li, H.; Liu, C.-T. On the Material and Temperature Impacts of Interior Permanent Magnet Machine for Electric Vehicle Applications. *IEEE Trans. Magn.* **2008**, *41*, 4329–4332. [CrossRef]
2. Fernando, N.; Vakil, G.; Arumugam, P.; Amankwah, E.; Gerada, C.; Bozhko, S. Impact of Soft Magnetic Material on Design of High-Speed Permanent-Magnet Machines. *IEEE Trans. Ind. Electron.* **2016**, *64*, 2415–2423. [CrossRef]
3. Al-Qarni, A.; El-Refaie, A. Design and Optimization of High-Performance Rare-Earth Free Interior Permanent Magnet Motors for Electric Vehicles Enabled by Iron Nitride Magnets. In Proceedings of the 2023 IEEE Energy Conversion Congress and Exposition (ECCE), Nashville, TN, USA, 29 October–2 November 2023; pp. 4082–4089.
4. Lian, G.; Li, H.; Chen, B.; Ban, F.; Zhang, J. Characteristic Analysis and Temperature Rise Calculation of PMSM Under Different Power Supply Modes. *IEEE Trans. Appl. Supercond.* **2019**, *29*, 1–5. [CrossRef]
5. Woo, J.H.; Bang, T.K.; Lee, H.K.; Kim, K.H.; Shin, S.H.; Choi, J.Y. Electromagnetic Characteristic Analysis of High-Speed Motors with Rare-Earth and Ferrite Permanent Magnets Considering Current Harmonics. *IEEE Trans. Magn.* **2021**, *57*, 8201805. [CrossRef]
6. Shrivatsaan, M.M.; GB, N.; Palka, R.; Wardach, M.; Prajzendanc, P.; Gundabattini, E.; Singh, R.R.; Gnanaraj, S.D. Performance analysis of electric motors, comparing lightweight techniques and cooling methods: A review. *Proc. Inst. Mech. Eng. Part C J. Mech. Eng. Sci.* **2023**, *238*, 1732–1746. [CrossRef]
7. Arita, H.; Nishiwaki, K.; Iezawa, M.K.T. High Speed Homopolar Type Permanent Magnet Motor Designed for Fast Ac-celeration Response. *IEEJ J. Ind. Appl.* **2022**, *11*, 763–770.
8. Lee, J.J.; Lee, K.D.; Kim, H.J.; Yoon, M.H. Design of Concentrated Flux Synchronous Motor with Ferrite Magnet Considering Me-chanical Stress at High Speed. *J. Magn.* **2021**, *26*, 166–174. [CrossRef]
9. Wang, Q.; Wang, C.; Li, Y. Performance evaluating of rotor arrangement on high torque density in-wheel permanent magnet machines. *IET Electr. Power Appl.* **2023**, *17*, 813–823. [CrossRef]

10. Jung, S.; Ko, J. Study on Regenerative Energy Recovery of Electric Vehicle through Voltage Control Using Switched Ca-pacitor. *IEEE Trans. Veh. Technol.* **2021**, *70*, 4324–4339. [CrossRef]
11. Hamednia, A.; Murgovski, N.; Fredriksson, J.; Forsman, J.; Pourabdollah, M.; Larsson, V. Optimal Thermal Management, Charging, and Eco-Driving of Battery Electric Vehicles. *IEEE Trans. Veh. Technol.* **2023**, *72*, 7265–7278. [CrossRef]

*World Electric Vehicle Journal*

*Article*

# A Novel Battery Temperature-Locking Method Based on Self-Heating Implemented with an Original Driving Circuit While Electric Vehicle Driving: A Numerical Investigation

Wei Li [1,2], Shusheng Xiong [2,3,*] and Wei Shi [3]

[1] Jiaxing Research Institute, Zhejiang University, Jiaxing 314031, China; liweisail@126.com
[2] Longquan Industrial Innovation Research Institute, Longquan 323700, China
[3] College of Energy Engineering, Zhejiang University, Hangzhou 310027, China; shiw@zju.edu.cn
* Correspondence: xiongss@zju.edu.cn

**Citation:** Li, W.; Xiong, S.; Shi, W. A Novel Battery Temperature-Locking Method Based on Self-Heating Implemented with an Original Driving Circuit While Electric Vehicle Driving: A Numerical Investigation. *World Electr. Veh. J.* **2024**, *15*, 408. https://doi.org/10.3390/wevj15090408

Academic Editor: Joeri Van Mierlo

Received: 29 July 2024
Revised: 29 August 2024
Accepted: 4 September 2024
Published: 6 September 2024



**Abstract:** In extremely cold environments, when battery electric vehicles (BEVs) are navigating urban roads at low speeds, the limited heating capacity of the on-board heat pump system and positive temperature coefficient (PTC) device can lead to an inevitable decline in battery temperature, potentially falling below its permissible operating range. This situation can subsequently result in vehicle malfunctions and, in severe cases, traffic accidents. Henceforth, a novel battery self-heating method during driving is proposed to maintain battery temperature. This approach is ingeniously embedded within the heating mechanism within the motor driving system without any necessity to alter or modify the existing driving circuitry. In the meantime, the battery voltage can be regulated to prevent it from surpassing the limit, thereby ensuring the battery's safety. This method introduces the dead zone into the space vector pulse width modulation (SVPWM) algorithm to form the newly proposed dSVPWM algorithm, which successfully changes the direction of the bus current in a PWM period and forms AC, and the amplitude of the battery alternating current (AC) can also be controlled by adjusting the heating intensity defined by the ratio of the dead zone and the compensation vector to the original zero vector. Through the Simulink model of the motor driving system, the temperature hysteresis locking strategy, grounded in the field-oriented control (FOC) method and employing the dSVPWM algorithm, has been confirmed to provide controllable and sufficiently stable motor speed regulation. During the low-speed phase of the China Light Vehicle Test Cycle (CLTC), the battery temperature fluctuation is meticulously maintained within a range of ±0.2 °C. The battery's minimum temperature has been successfully locked at around −10 °C. In contrast, the battery temperature would decrease by a significant 1.44 °C per minute without the implementation of the temperature-locking strategy. The voltage of the battery pack is always regulated within the range of 255~378 V. It remains within the specified upper and lower thresholds. The battery voltage overrun can be effectively avoided.

**Keywords:** lithium-ion battery; extremely cold environment; internal self-heating; motor drive system

## 1. Introduction

In scenarios where battery electric vehicles (BEVs) are started and operated in exceptionally frigid conditions of −30 °C or colder, particularly when navigating urban roads at slow speeds, it is inevitable that, despite the battery's initial discharge at a minimal current, the overall battery temperature will decline or potentially drop beneath its acceptable operating threshold. At this time, the battery polarization increases, the voltage drops rapidly, and for the sake of battery life and safety, the battery management system (BMS) will limit the battery's ability to continue working, causing the vehicle to break down and even causing traffic accidents. Therefore, the battery must be heated while driving to restore its state of power (SOP). There are two main types of existing on-board heating technology solutions: the first is heat pump air conditioning. Lee et al. [1] showed that at an ambient temperature of 10 °C, the coefficient of performance (COP) of the vehicular heat pump can reach 2.4. However, at an ambient temperature of $-10$ °C, a heat pump using a single heat source cannot work effectively, and the COP is close to 1. Heat pumps that can work in extremely low-temperature environments (below $-20$ °C) currently do not have a good industrial solution [2]. The second is adopting a heater composed of positive temperature coefficient (PTC) material, or an alternative dissipative resistor, to directly heat the circulating coolant, which in turn heats the battery. However, when the temperature drops, the battery frequently lacks sufficient state of power (SOP) to cater to the full power operation of the heater, which is around 4 kW in the case of mid-range passenger cars. Consequently, this approach to heating proves to be excessively sluggish or even impractical.

AC internal heating is a technology that relies on the internal materials of the battery to generate heat under the action of alternating current, which has the advantages of fast and homogeneous temperature rise, less heat loss to the surroundings, and no heating element is needed [3]. The increase in impedance of Lithium-ion (Li-ion) batteries at low temperatures is to the benefit of battery self-heating because, under the same amplitude of alternating current (AC), a larger amount of ohmic and electrochemical heat is generated inside the battery [4]. In the past ten years, many scholars have studied the low-temperature heating characteristics of batteries through AC power cabinets in the laboratory [3,5–12]. These research results verified the feasibility of battery self-heating technology. More importantly, these studies indicate that if the parameters of alternating current (amplitude and frequency) are properly selected to keep the battery terminal voltage within the upper and lower limits, heating lithium-ion batteries at extremely low temperatures using AC heating has been proven to be effective and benign. For instance, Ruan H et al. concluded that as long as the battery terminal voltage is controlled within its upper and lower limits, there will be almost no significant battery life degradation in the short term [13]. Guo S et al. studied the aging effects of repeated AC heating cycles on lithium-ion batteries at $-20$ °C through capacity calibration and incremental capacity analysis. These heating processes were always without overvoltage. The results showed no adverse effects on the cell's lifespan [14]. Additionally, the studies by Zhu J [3] and Jiang J [10] also reached the same conclusion. However, the above studies were realized under laboratory conditions and cannot be applied to the vehicle environment. Research in recent years has focused on how battery self-heating methods can be applied to vehicles. One scheme is to additionally design a separate part of the vehicle that is dedicated to battery self-heating [15–22]. Undoubtedly, this pattern increases the cost and weight of the vehicles. The other pattern is to integrate battery heating with a vehicle subsystem, in which the most promising method is to embed the heating strategy in the motor driving system. Most of these methods require changes to the topology of high-voltage wiring and are only applicable to newly developed vehicles [23–26]. In 2022, Du C. et al. proposed a battery internal heating method without any additional hardware modification in a cold environment [27]. However, this method can only be used for parking heating. So far, no research has been found on the battery self-heating method that can be used in the driving process without changing the original circuit topology of the vehicle, nor on the battery temperature control strategy with adjustable heating intensity.

To solve the problem of battery temperature drop during the driving of BEVs in extremely cold environments and fill the research gap in this direction, this paper proposes a new method for locking battery temperature during driving. The proposed method is entirely rooted in the existing motor driving system, mandating neither the integration of additional hardware nor alterations to the original circuitry; it aims to achieve battery self-heating exclusively by finely tuning control strategies. In consideration of the circuitry's inherent compatibility and suitability, it is expected to yield no adverse effects. By modifying the underlying strategy for motor control, the battery temperature is held higher than the set threshold, and the battery voltage is controlled so that it does not exceed the limit to ensure the safety of the battery. Thus, customers can obtain an enhanced driving experience, including an extended driving range in winter, with virtually no extra cost incurred from the vehicle purchase. The content of this paper mainly includes the following: firstly, the theoretical basis of the permanent magnet synchronous motor (PMSM) and its field-oriented control (FOC) are analyzed. Based on that, with the introduction of the dead-zone insertion method, the space vector pulse width modulation (SVPWM) mechanism is improved, and the PWM waveform that can realize the AC battery current is designed. A control strategy for regulating AC battery current to lock the battery temperature and avoid the battery voltage overrun is further designed. Secondly, the effectiveness of the model and the method was preliminarily verified by small-scale bench experiments. Through simulation, the precision, dynamic performance, and stability of the control strategy are verified compared with the original PMSM control algorithm. Finally, through the simulation under the China light vehicle test cycle (CLTC), the effect of the proposed method in real vehicle driving conditions is further evaluated. The contribution of this paper consists of (1) the proposed method that eliminates the necessity for altering the existing circuit topology or incorporating additional hardware in the BEV, and it remains compatible with the current PMSM FOC algorithm; (2) a novel mechanism that has been clearly identified for generating AC in the battery via the motor drive system while the vehicle is cruising; and (3) a unified computing platform that has been devised to integrate seamlessly the battery, motor, and controller, thereby enabling deeper insights into the battery self-heating mechanisms integrated with the motor drive system. This development paves the way for further exploration and refinement of the system's battery thermal management capabilities.

## 2. Material and Methods

### 2.1. The Introduction of Dead Zones

The FOC technology developed in the 1980s of the 20th century is now widely used in the control of PMSMs in BEVs. The hardware of a conventional three-phase PMSM inverter exhibits a three-phase full-bridge topology consisting of 6 power electronic devices. Power metal–oxide–silicon field-effect transistors (MOSFET) are employed, and the typical circuit topology is shown in Figure 1a. The three-phase stator windings of the motor are simplified to three inductors, A, B, and C, connected in a Y-type. The half-bridge consisting of two MOSFETs connected to each inductor is correspondingly called the A-phase, B-phase, and C-phase half-bridge. MOSFETs connected to the positive terminal of the power source are defined as upper tubes, which are represented as $T_{AU}$, $T_{BU}$, and $T_{CU}$, respectively, and the rest are called down tubes, which are represented as $T_{AD}$, $T_{BD}$, and $T_{CD}$, respectively. The power source has a voltage defined as $V_{dc}$.

The midpoint voltage $V_i$ ($i$ = A, B, C) of each half-bridge depends on the switching state of the MOSFETs, which constitute the half-bridge itself. The switch states are defined as $S_i$ ($i$ = A, B, C), assigned to 0 or 1, and their correspondence to midpoint voltage is shown in Table 1.

These different combinations of switch states form the basic voltage vector. The voltage vector shown in Figure 1b contains eight basic vectors consisting of 1 and 0, six of which are non-zero vectors or valid vectors: $U_1 \sim U_6$, and two zero vectors: $U_0$ and $U_7$. Six non-zero vectors equitably divide an electrical period into six regions: I~VI in space. Through these eight basic vectors, spatial voltage vectors of any direction and size can be synthesized to form a rotating stator magnetic field. It is noteworthy that the time slice in which the upper and lower tubes are simultaneously turned off is defined as the dead zone, $U_x$, and is not included in the basic vectors.

In the FOC strategy, when the motor is in a driving condition, the direction of the bus current is always positive, and the battery is discharged. How to change the direction of the bus current and quickly charge and discharge the battery is the key issue to achieving battery AC self-heating. For this purpose, a dead zone is introduced to turn off the upper and down tubes of a phase simultaneously, and then the value of $V_i$ is uncertain, which is opposite to the midpoint voltage of the basic vector before the dead zone. This makes it possible to reverse the direction of the bus current.

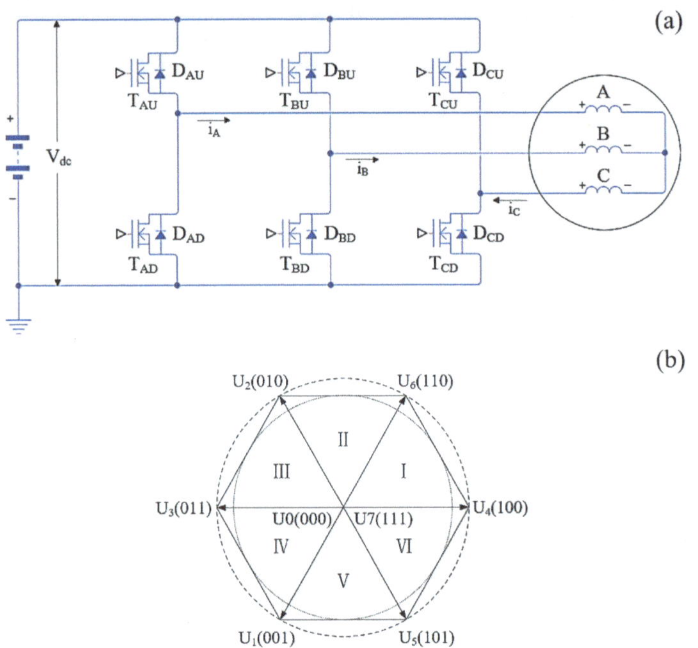

**Figure 1.** Principle of FOC of three-phase PMSM: (**a**) circuit topology of the inverter; (**b**) basic voltage vector diagram.

**Table 1.** Definition of switch states and their corresponding midpoint voltage.

| State of $T_{iU}$ * | State of $T_{iD}$ * | Value of $S_i$ * | Value of $V_i$ * |
| :---: | :---: | :---: | :---: |
| ON | OFF | 1 | $V_{dc}$ |
| OFF | ON | 0 | 0 |
| OFF | OFF | x | $V_{dc}$ or 0 |

* Subscript character i represents A, B, or C.

### 2.2. The Effect of Dead Zones

Taking the vector $U_4$ as an example, the determining principle of $V_i$ in dead zones and bus current is illustrated, as shown in Figure 2.

When $T_{AU}$ is on, $V_A$ is equal to $V_{dc} - V_{sw}$, where $V_{sw}$ is the on-voltage drop of the MOSFET. Since $T_{BD}$ and $T_{CD}$ are on and their upper MOSFETs are closed, $V_B$ and $V_C$ are $V_{sw}$. At this point, the current in phase A, $i_A$, increases continuously in the direction shown by the green dotted arrow in Figure 2a. The sum of $i_B$ and $i_C$ is equal to $i_A$, and $i_A$ equals the battery current. If the inverter is switched to the dead zone at this time, a turn-off gating signal is sent to all of the MOSFETs. The current in $T_{AU}$ gradually decreases during its device storage time, and the corresponding current is transferred to the body diode $D_{AD}$ at the down tube of phase A because $i_A$ needs to be maintained. The voltage across $T_{AU}$ rises, and the voltage across the diode $D_{AD}$ drops to its on-voltage $V_d$. Thus, $V_A$ is reduced to $-V_d$. Similarly, since the MOSFETs are all off, the current in $T_{BD}$ and $T_{CD}$ will also be transferred to the corresponding body diodes. The direction of the three-phase current is shown by the orange dashed arrow in Figure 2b. The current of phase A during the dead zone, $i_A'$, is equal to the battery current. It can be seen that switching from a non-zero vector to a dead zone can realize the commutation of the battery current. In general, when the non-zero vector switches to the dead zone, the battery current reverses, the midpoint voltages Vi is transformed into the opposite of the original value, and the motor phase voltages are also inverse. Similarly, when switching from a dead zone to one of the non-zero vectors, the conclusion is the same.

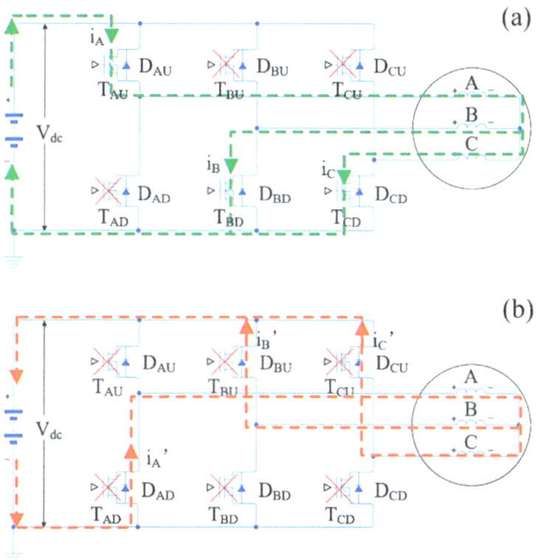

**Figure 2.** Schematic of the reverse current in FOC: (**a**) the current of each phase at vector U1; (**b**) the current of each phase in the dead zone.

The presence of a DC-link capacitor will weaken the amplitude of AC current, especially when the heating frequency is close to the resonance point of the resistor and inductor paralleled with the capacitor (RL-C) of the DC-link capacitor and motor stator circuit. Details can be referenced in the author's previous early-access article [28]. However, this situation can be avoided in application, and thus, the impact of the DC-link capacitor can be ignored. Therefore, if the influence of the DC-link capacitance is not considered, the correspondence between the bus current $i_{bat}$ and the phase current can be referred to in Table 2. The battery current is equal to the inverter bus current, which is directly related to the heating power of the battery. Then, the state of charge (SOC) of the battery can be further calculated.

**Table 2.** The correspondence between the bus current $i_{bat}$ and the phase current $i_i$ ($i$ = A, B, C) during different vectors.

| Vectors | $S_A$ | $S_B$ | $S_C$ | $i_{bat}$ |
|---|---|---|---|---|
| $U_4$ | 1 | 0 | 0 | $i_A$ |
| $U_6$ | 1 | 1 | 0 | $-i_C$ |
| $U_2$ | 0 | 1 | 0 | $i_B$ |
| $U_3$ | 0 | 1 | 1 | $-i_A$ |
| $U_1$ | 0 | 0 | 1 | $i_C$ |
| $U_5$ | 1 | 0 | 1 | $-i_B$ |
| $U_{nX}$ * | x | x | x | $-\max(|i_i|)$ * |

* Subscript character i represents A, B, and C, and n represents 1~6.

Table 2 manifests that the magnitude of the bus current during the dead zone depends on the three-phase current of the previous non-zero vector and is numerically equal to the opposite of the largest absolute values in the three-phase current.

From the vector diagram in Figure 1b, the dead zone after each non-zero vector is the same as the vector with a phase difference of 180° from the non-zero vector itself. However, the switch combination is completely different. Under the condition of equal magnitude transformation, the base vector in the vector diagram is determined by the following equation:

$$\mathbf{V}_{out} = \frac{2}{3}V_{dc}\left(s_A + s_B e^{j\frac{2\pi}{3}} + s_C e^{j-\frac{2\pi}{3}}\right) \tag{1}$$

Thus, the dead-zone vector corresponding to the base vector i can be expressed as:

$$\mathbf{V}_{out\,nX} = \mathbf{V}_{out\,n} e^{j\pi}(n = 1 \sim 6) \tag{2}$$

It can be seen that the output voltage vector $\mathbf{V}_{out\,nX}$ of any dead zone can always find a non-zero vector $\mathbf{V}_{out\,n}$, rendering the modulo of the two equal and the phase difference 180°. In the real FOC algorithm, the time control of the dead zone is particularly important. If the dead zone is too long, the current shown in Figure 2b will eventually decay to 0, all body diodes will be off, and the output voltage vector of the dead zone will no longer be equal to the reverse vector of the last vector but will be equivalent to the zero vector. When the zero vector is applied, the phase voltage and line voltage of the motor are all zero, so the phase current decays rapidly with time. The action time of the zero vector is uncertain in FOC and is related to the PWM period and output power. Although the zero vector cannot realize the commutation of the bus current, it is of great significance in regulating the heating power inside the battery.

### 2.3. Heating Intensity Adjustment Mechanism

To synthesize a space-rotating magnetic field, the SVPWM algorithm is used. The theoretical basis of the SVPWM algorithm is the principle of mean equivalence. That is, within a PWM period $T_s$, the basic voltage vectors are arranged so that their average value is equal to the given voltage vector. The output voltage vector synthesis, considering the dead zone, still needs to comply with the above principle in the natural coordinate system. Taking region I as an example, at some point, the output voltage vector $\mathbf{V}_{out}$ is synthesized by $U_4{}'$ and $U_6{}'$ in the direction of two non-zero base vectors of that region. To form the alternating current, the corresponding dead-zone vectors $U_{4X}$ and $U_{6X}$ are introduced, as shown by the red arrows in Figure 3. Since the dead-zone vector is inverse to its corresponding base vector, it is equivalent to reducing the magnitude of the original base vector. Therefore, in order to ensure the magnitude of the output voltage vector, the compensation vectors $U_{4c}$ and $U_{6c}$ should be introduced, which have the same direction as their base vectors.

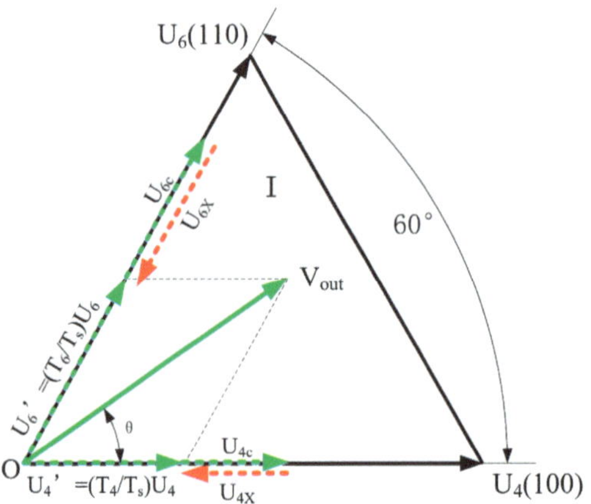


**Figure 3.** A spatial vector composition method with dead zones.

From the figure above, the compensation vector and the dead-zone vector seem to cancel each other geometrically and do not affect the direction and magnitude of the output voltage vector. Both the compensation vector and the dead-zone vector need to occupy the action time of SVPWM, which will affect the magnitude of the output voltage vector. From this point of view, the superposition of the dead-zone vector and the compensation vector has the same effect on the output voltage vector as the zero vector. Therefore, this paper intends to modify the conventional SVPWM algorithm to replace part of the zero vector with a dead-zone vector and a compensation vector, hereinafter referred to as the dead-zone space vector pulse width modulation (dSVPWM) algorithm. According to the above analysis, the dSVPWM algorithm needs to meet the following equations:

$$T_s \mathbf{V_{out}} = T_4 \mathbf{U_4} + T_6 \mathbf{U_6} \tag{3}$$

$$T_s = T_4 + T_6 + T_0 \tag{4}$$

$$T_0 = T_0' + T_{4X} + T_{6X} + T_{4c} + T_{6c} \tag{5}$$

where $T_0$ represents the action time of the zero vector in the SVPWM algorithm, and $T_0'$ represents the zero vector retained by dSVPWM. $T_4$ and $T_6$ represent the action time of non-zero base vectors $U_4$ and $U_6$, $T_{4X}$ and $T_{6X}$ represent the action time of the dead-zone vectors $U_{4X}$ and $U_{6X}$ in the dSVPWM algorithm, and $T_{4c}$ and $T_{6c}$ represent the action time of the compensation vectors $U_{4c}$ and $U_{6c}$, respectively. Theoretically, $T_{4X}$ equals $T_{4c}$, and $T_{6X}$ equals $T_{6c}$. However, considering the nonlinearity of the circuit components, the compensation time needs to be adjusted appropriately. The ratio of the compensation time and dead time is defined as $a_{cX}$, hereinafter referred to as the compensation coefficient, and its expression is as follows:

$$a_{cx} = \frac{T_{6c}}{T_{6X}} = \frac{T_{4c}}{T_{4X}} \tag{6}$$

The compensation coefficient $a_{cX}$ should be set to around 1; otherwise, the magnitude of $U_4'$ and $U_6'$ will be changed. In practice, calibration experiments for $a_{cX}$ can be performed according to the phase current waveform. The ratio of $T_{4X}$ and $T_{6X}$ does not theoretically affect the output voltage vector. So, to make the equations have a unique solution, the following equation is specified:

$$\frac{T_{4X}}{T_{6X}} = \frac{T_4}{T_6} \tag{7}$$

In addition, define the ratio of time containing the dead zone and the compensation vector to the time of the original zero vector as heating intensity with the symbol $b_n$:

$$b_n = \frac{T_0 - T_0'}{T_0} = \frac{T_{4x} + T_{6x} + T_{4c} + T_{6c}}{T_0' + T_{4x} + T_{6x} + T_{4c} + T_{6c}} \tag{8}$$

The value of $b_n$ indicates the time proportion of the AC production composed of the dead zone and its compensation vector. It can be predicted that the larger the $b_n$, the greater the AC of the battery. Therefore, the intensity of battery AC self-heating can be controlled by adjusting $b_n$.

So far, by inputting the magnitude $V_{out}$ and direction $\theta$ of any target voltage vector and fixing the values of $a_{cX}$ and $b_n$, $T_4$ and $T_6$ can be calculated according to Equations (9)–(11) and then the action time of each dead-zone vector and compensation vector can be obtained according to Equations (4)–(8).

$$T_4 = \frac{V_{out} T_s}{U_4} \left( \cos\theta - \frac{\sqrt{3}}{3} \sin\theta \right) \tag{9}$$

$$T_6 = \frac{2\sqrt{3}}{3}\frac{V_{out}T_s}{U_6}\sin\theta \tag{10}$$

$$U_4 = U_6 = \frac{2}{3}V_{dc} \tag{11}$$

The physical interpretation can be obtained by multiplying each solved time by the corresponding vector, which means that the volt-second area obtained by the output voltage vector $\mathbf{V}_{out}$ acting on one PWM period Ts can be obtained by summing the base vectors $U_4{}'$ and $U_6{}'$, the dead-zone vectors $U_{4X}$ and $U_{6X}$, and the compensation vectors $U_{4c}$ and $U_{6c}$ multiplying with their corresponding action times. The magnitude of these vectors and their action time can be converted into a time series of switching the MOSFETs, which form corresponding PWM waves. The calculation method of the action time in other sectors is the same as in sector I.

The modulation ratio of dSVPWM is defined as:

$$M = \frac{\sqrt{3}|\mathbf{V}_{out}|}{V_{dc}} \tag{12}$$

To ensure that the composition vector is inside the linear region of a regular hexagonal, as shown in Figure 1b, the maximum magnitude of the output voltage vector $\mathbf{V}_{out}$ must be satisfied by $|\mathbf{V}_{out}| \leq 2V_{dc}/3$, while dSVPWM does not insert new non-zero vectors during each PWM period. Therefore, the maximum modulation ratio can still be 1.1547, which is consistent with the SVPWM algorithm.

After obtaining the action time of each vector, as long as the wave order is determined, the waveform of dSVPWM can be plotted. In this paper, a PWM scheme, as shown in Figure 4, is designed. Considering the adjustability for the time of the dead-zone vector, part of the zero vector is retained. To reduce the current ripple, a double-pulse mode is used per cycle. Each non-zero vector is divided equally into two parts and placed separately in each half period to maintain the symmetrical waveform pattern, thereby effectively reducing the harmonic component of PWM. From the analysis of Figure 4, it can also be known that the battery current alternates between positive and negative within a PWM cycle: during the non-zero vector action time slice, the battery current is negative, and the battery discharges; during the dead-zone slice, the battery current is positive, and the battery charges. Therefore, the period of the battery AC must be consistent with the PWM period. It is unrelated to the period of the phase current and the poles of the motor. This phenomenon will also be illustrated with an example in Section 4.2.2.

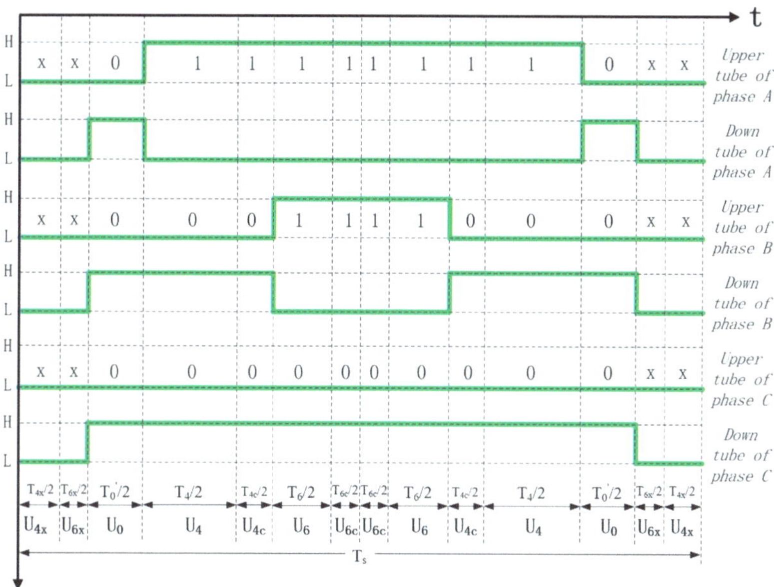


**Figure 4.** Switching waveform and vector composition of dSVPWM in one period.

To reduce the times of switching, each dead-zone vector and the compensation vector are also split in half so that the dead-zone vector is located at both ends to maintain the symmetry of the waveform and minimize the times of switching action. Each time toggling from a zero vector to a dead-zone vector, 3 MOSFETs need to turn off, and toggling from a dead-zone vector to a zero vector requires 3 MOSFETs to turn on. Therefore, the times of MOSFET switching per cycle is 14. Compared with the 12 times of the SVPWM algorithm based on the software mode, this is an increase of 2 times. For MOSFET inverters, switching losses are not the main source of power loss, and the small increase in the times of switching does not have much effect.

The general equation of battery heating power can be obtained from Joule's law:

$$P_{bat} = i_{bat\_s}^2 R_{bat} \tag{13}$$

where $R_{bat}$ represents the equivalent ohmic internal resistance of the battery, $P_{bat}$ represents the battery heating power, and $i_{bat\_s}$ represents the root-mean-square value of the battery current in one FOC period. According to the dSVPWM algorithm proposed above, taking region I as an example, when $i_A > 0$, $i_B < 0$, and $i_C < 0$, the quadratic of the $i_{bat\_s}$ can be expressed as:

$$i_{bat\_s}^2 = \frac{1}{T_s} \left( \int_0^{(T_4+T_{4c})/2} i_A^2 dt + \int_0^{T_6+T_{6c}} i_C^2 dt + \int_0^{(T_4+T_{4c})/2} i_A^2 dt + \int_0^{T_{4x}+T_{6x}} i_A^2 dt \right) \tag{14}$$

From Equation (14), it can be seen that the battery heating power is not only related to the value of the motor phase current but also related to its action time in a FOC period. It is noteworthy that in the traditional SVPWM scheme, the battery current is zero when the zero vector is applied, while in the proposed dSVPWM algorithm, in the primordial zero vector action time, an alternating battery current is formed, which effectively improves the battery heating power. Moreover, under the action of the dead-zone vector, the absolute value of the battery current always takes the maximum value of the three-phase current, as shown in Table 2, thus further increasing the battery heating power.

On the other hand, the proposed dSVPWM changes the ripple of the phase current, which directly determines the electromagnetic force of the motor. This will inevitably produce high-frequency torque fluctuations on the motor output shaft and may form high-frequency noise. The torque fluctuations will be discussed in Section 4.2.5 in the following text, while NVH is not the focus of this paper and will be analyzed in subsequent research.

*2.4. Small-Scale Motor Driving System Test Bench*

A small-scale motor driving system test bench, as shown in Figure 5, was built to verify the feasibility and advantages of the above heating strategy. The battery pack was assembled by 12 cells (INR18650/25P, EVE Energy Co., Ltd., Huizhou, China) connected in series. Its rated voltage is 43.2 V, and the maximum discharge rate is 10 C. The battery pack was fully charged to 2.5 Ah at room temperature and discharged to a SOC of 60%. The surface of the battery pack was not insulated and was in natural convection with the ambient air. The terminals of the battery pack were connected to the homemade driver board through the direct current (DC) link, and two wires were connected to the control board to accurately measure the terminal voltage of the battery pack. The DC-link currents and battery voltages were monitored and recorded by an oscilloscope (DSO2014A, Keysight, Santa Rosa, CA, USA) during the test. The motor used in the test is a PMSM for an electric tricycle with a rated power of 1 kW. Its detailed parameters are shown in Table 3. These parameters are provided by the vehicle manufacturer. The motor was mounted on a 2 kW electromagnetic dynamometer (ZD10, Guangzhong Electrical Equipment Co., Ltd., Taizhou, China). Forced air cooling was adopted for the controller and inverter by a fan to prevent overheating. The motor was cooled by natural convection at room temperature. A digital signal processor (DSP) is embedded in the control board, which is used to run the control strategy designed in this paper.

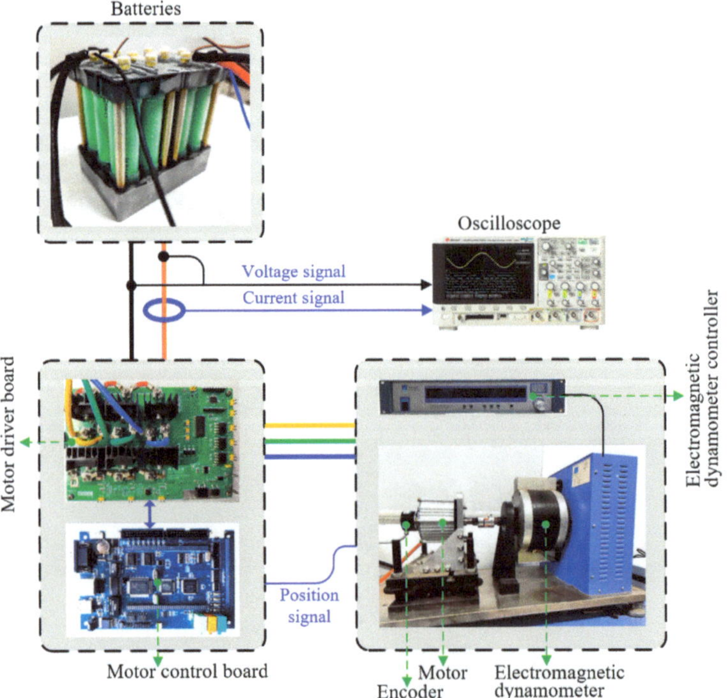


**Figure 5.** Small-scale motor driving system test bench.

**Table 3.** Parameters of the motor in the test.

| Parameters | Value | Unit |
|---|---|---|
| Rated current | 27 | A |
| Rated voltage | 48 | V |
| Inductance of d-axis | 71.2 | uH |
| Inductance of q-axis | 99.5 | uH |
| Phase resistance | 15.1 | m$\Omega$ |
| Pole pairs | 5 | - |
| Flux linkage | $1.67 \times 10^{-2}$ | Wb |
| Back EMF constant | 15.4 | V/krpm |
| Rotational inertia | $1.90 \times 10^{-4}$ | kg·m$^2$ |
| Friction coefficient | $3.20 \times 10^{-3}$ | N·m·s |

## 3. Simulation and Calculation

### 3.1. Small-Scale System Models for Experimental Validation

To verify whether the dSVPWM algorithm can be used in the FOC control strategy of PMSM and realize AC heating of the battery while driving, based on the algorithm described in Section 2, the physical model of the small-scale motor driving system test bench is built in Simulink to carry out the numerical calculation. The parameters used in the model are the same as those on the test bench. The battery is programmed by Matlab function as a fractional order equivalent circuit model, and its parameters are measured by the electrochemical workstation (CS1350, Corrtest Instruments Co., Ltd., Wuhan, China). Firstly, the precision of the model is verified by comparing it with the small-scale test bench, and then the characteristics of the proposed algorithm are studied by comparing the dSVPWM algorithm with the original SVPWM algorithm. The model adopts a closed-loop structure with a dual PI of current and speed. The PI loop for current employs internal mode control for PI parameter tuning [29]. PI parameter tuning of the speed loop is based on the active damping method proposed in the literature [30]. The completed model is shown in Figure 6. The blue shading module contains the dSVPWM algorithm. The green shading module is the original SVPWM algorithm for comparative studies. The controller in the model adopts the basic FOC method, id = 0 control, and the parameters used are shown in Table 4.

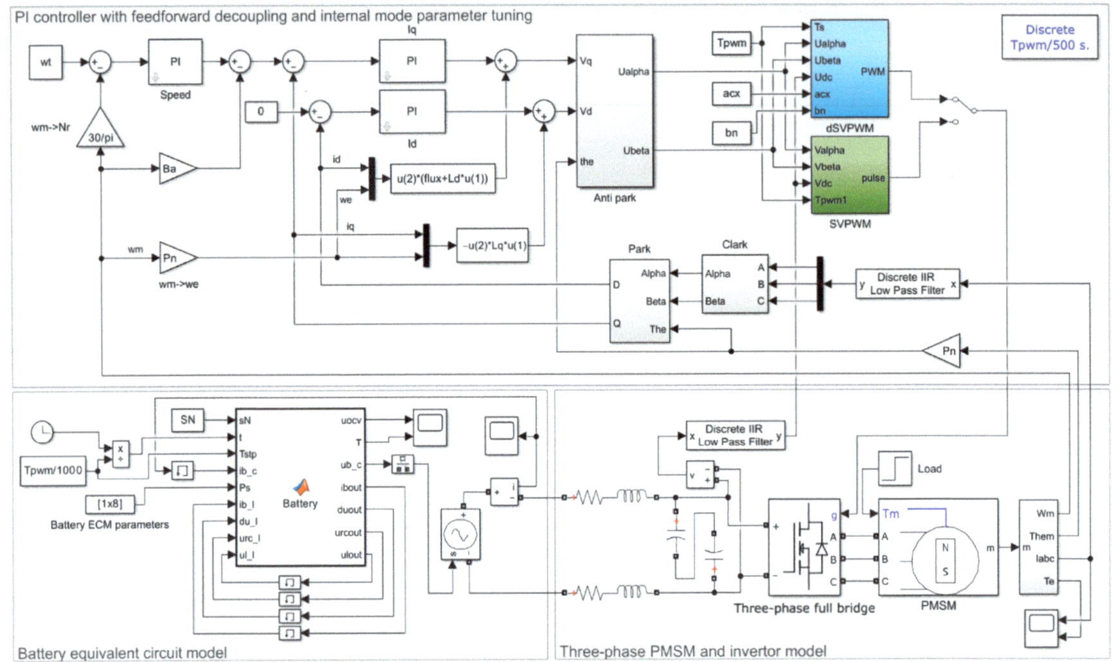


**Figure 6.** Powertrain models of a BEV in Simulink.

**Table 4.** Parameters of the controller in the model.

| Parameters | Value | Unit |
| --- | --- | --- |
| Proportional gain for d-axial current loop | $3.56 \times 10^{-2}$ | - |
| Integral gain for d-axial current loop | 7.55 | - |
| Proportional gain for q-axial current loop | $4.98 \times 10^{-2}$ | - |
| Integral gain for q-axial current loop | 7.55 | - |
| Proportional gain for speed loop | $7.6 \times 10^{-2}$ | - |
| Integral gain for speed loop | 3.8 | - |
| PWM period | $1 \times 10^{-4}$ | s |
| PI controller period | $1 \times 10^{-3}$ | s |
| Bandwidth for speed loop | 50 | rad/s |
| Bandwidth for the current loop | 2500 | rad/s |

### 3.2. Vehicle-Level Powertrain Model and Parameters

In order to verify the proposed method for locking battery temperature while driving at the vehicle level, the parameters in the model shown in Figure 6 are refreshed according to the motor and battery pack parameters of a pure electric passenger car reported in the literature [27]. The main parameters used in the vehicle-level model are shown in Tables 5 and 6.

**Table 5.** Parameters of the motor in the vehicle-level model.

| Parameters | Value | Unit |
| --- | --- | --- |
| Peak power | 160 | kW |
| Rated voltage | 354 | V |
| Inductance of d-axis | 0.09 | mH |
| Inductance of q-axis | 0.24 | mH |
| Phase resistance | 5 | m$\Omega$ |
| Pole pairs | 4 | - |
| Flux linkage | 0.036 | Wb |

**Table 6.** Parameters of the battery pack in the vehicle-level model.

| Parameters | Value | Unit |
| --- | --- | --- |
| Initial temperature | −10 | °C |
| Ambient temperature | −35 | °C |
| Initial SOC | 65 | % |
| Serial number | 90 | s |
| Rated voltage | 324 | V |
| Nominal capacity | 100 | Ah |
| DC-link Equivalent resistance | 0.5 | m$\Omega$ |
| DC-link Equivalent inductance | 8 | μH |
| Specific heat capacity | 935 | $J \cdot kg^{-1} \cdot K^{-1}$ |
| Surface heat transfer coefficient | 10 | $W \cdot m^{-2} \cdot K^{-1}$ |
| Thermal conductivity in axial | 15.1 | $W \cdot m^{-1} \cdot K^{-1}$ |
| Thermal conductivity in radial | 0.45 | $W \cdot m^{-1} \cdot K^{-1}$ |

The equivalent circuit model of the cell remains the same as the model described in Section 3.1, and after the cell is properly connected in series and parallel, the voltage and capacity required for the motor in Table 5 are obtained. In addition, the lumped-parameter heat transfer model of the cell is added here to calculate the temperature rise of the battery. The heating power of the cell is calculated by Equation (14). The thermophysical parameters are listed in Table 6.

The ambient temperature in the model was set to −35 °C to simulate extreme cold weather. The initial battery temperature is set to −10 °C, which represents the battery temperature of the vehicle after it has been warmed up during parking.

In the above Simulink model, let the motor run following the CLTC speed; meanwhile, calculate the battery voltage, current, temperature, and other parameters, which are used to evaluate whether the proposed temperature locking strategy can work in real operation of the vehicle, when it works, and how effective it is.

For this purpose, the speed of the vehicle's motor under CLTC and the corresponding load torque must be obtained. According to the longitudinal dynamic equation of the vehicle, the relationship between the driving force and the vehicle speed can be expressed as:

$$F_t = \delta m \frac{dv}{dt} + f_0 + f_1 v + f_2 v^2 \tag{15}$$

In the equation, $F_t$ represents the driving force. m is the vehicle test mass. $\delta$ is the rotational mass conversion coefficient. v represents the vehicle speed. $f_0$ and $f_1$ are the rolling resistance coefficients. $f_2$ is the coefficient related to wind resistance, which can be calculated as $C_D \cdot A \cdot v^2 / 21.15$, where $C_D$ is the vehicle's coefficient of air resistance and A is the windward area.

Combined with the parameters provided by a vehicle manufacturer, as shown in Table 7, the vehicle speed in the CLTC diagram can be converted into the speed of the motor, and the driving force can be calculated according to Equation (15), and the motor load torque can be further obtained. This paper only focuses on the data from 274 to 520 s in CLTC, which represents the characteristics of urban scenarios, such as low-speed driving, stop-and-go, etc. The vehicle speed before and after this period is much higher, and the battery temperature then has been verified as not dropping in simulation. The heat dissipation condition of the battery is set to natural convection with air. If the vehicle is driving, the heat dissipation conditions of the battery will be better, which means that the battery temperature drops more easily. Of course, the battery pack in BEVs now commonly uses liquid cooling or direct refrigerant cooling methods, in which the heat transfer process between the battery and the cooling medium is forced convection, and the heat exchange efficiency is much higher than that of natural convection. Therefore, the simulation conditions in this paper are relatively demanding, and the results must be more convincing.

**Table 7.** Parameters for calculating motor speed and torque from CLTC.

| Parameters | Value | Unit |
| --- | --- | --- |
| Vehicle test mass | 1990 | kg |
| Tire radius | 0.3588 | m |
| Rolling resistance coefficient $f_0$ | 133 | - |
| Rolling resistance coefficient $f_1$ | 1.097 | - |
| Coefficient $f_2$ | 0.041 | - |
| Rotational mass conversion coefficient | 1.018 | - |

The calculated motor speed and torque are shown in Figure 7. Obviously, vehicle speed is less than 20 km/h in the selected period, which includes three driving–stopping cycles: driving for 17 s then stopping for 35 s, driving for 26 s then parking for 50 s, and driving for 35 s then parking for 85 s. Acceleration and deceleration in the first two cycles are greater than those in the third one. This characteristic represents typical Chinese urban road conditions.

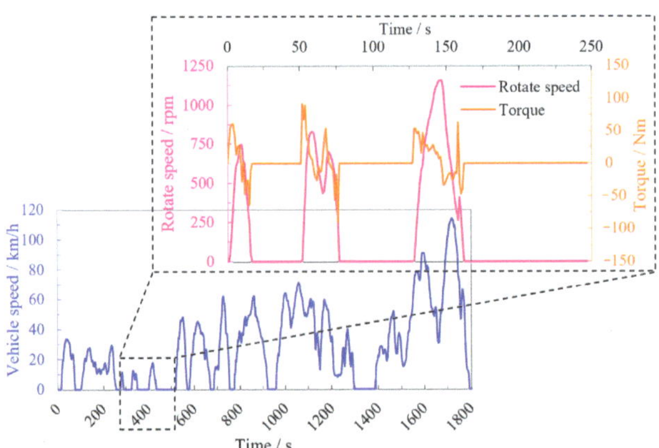


**Figure 7.** Motor speed and load torque calculated according to the low-speed period of CLTC.

### 3.3. Structure and Workflow of Hysteresis Control for Battery Temperature

Based on the proposed dSVPWM algorithm, this paper proposes a temperature-locking strategy for power batteries during vehicle driving, which employs hysteresis control to maintain the temperature of the battery within the set upper and lower thresholds and prevent the occurrence of the vehicle speed limit or accidental breakdown caused by the drop in battery temperature during driving.

The main idea of this strategy is to automatically adjust the motor control strategy according to the feedback of battery temperature and change the self-heating power of the battery by updating the battery heating intensity $b_n$ in real time so that the battery temperature is always maintained above the threshold. The structure of this control strategy is shown in Figure 8. The control period is the same as the PWM period. The target temperature consists of a low threshold $T_{bl}$ and a high threshold $T_{bh}$, the difference of which is the hysteresis band, which can be set to about 0.5 °C. In the model, $T_{bl}$ and $T_{bh}$ are set to $-10.3$ °C and $-9.8$ °C, respectively. When the average temperature of the battery is lower than $T_{bl}$, the dSVPWM algorithm is started, the battery heating intensity $b_n$ is increased in increments of $2 \times 10^{-4}$ per control period, and the change rate of the average temperature of the battery is monitored in real-time. If the change rate of battery temperature is still not inverted in the next control period, the heating intensity $b_n$ will continue to increase until the average battery temperature begins to rise. If the battery temperature reaches a high threshold $T_{bh}$, the dSVPWM algorithm will exit and revert to the SVPWM algorithm.

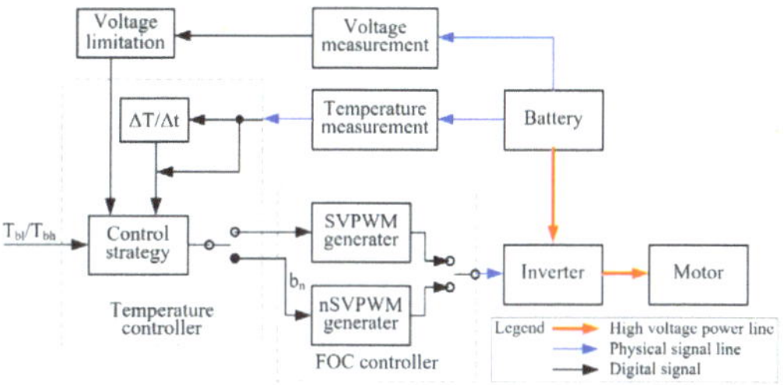


**Figure 8.** Control structure of hysteresis control for temperature locking during vehicle driving.

The algorithm flowchart is shown in Figure 9. In addition to the battery temperature control strategy, the figure also includes the control strategy for battery voltage. If the battery voltage is detected close to the upper or lower thresholds, the $b_n$ quickly decreases by a speed of $2 \times 10^{-3}$ per period to avoid voltage overruns due to the heating strategy. This is to avoid the battery terminal voltage exceeding its limited operating voltage at a large heating intensity when the battery SOC is high or low. If the battery voltage still exceeds the limit when the $b_n$ is reduced to zero, the security measures of the BMS will be triggered, which is outside the scope of this paper.

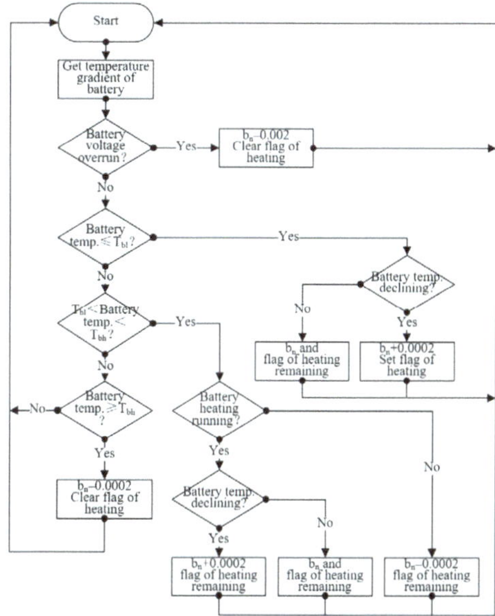


**Figure 9.** Flow chart of hysteresis control for temperature locking during vehicle driving.

## 4. Results and Discussion

### 4.1. Experimental Validation for Speed Control Based on dSVPWM

The model shown in Figure 6 was used to simulate the closed-loop control of the speed under fixed working conditions, and the synchronous comparison test was carried out on the small-scale electric drive system test bench shown in Figure 5. On the one hand, the effectiveness of the model structure and parameter was verified by the comparison test, and on the other hand, the advantages of the dSVPWM algorithm over the SVPWM algorithm in battery self-heating were also verified. These comparisons only concern battery current, as its amplitude and symmetry directly reflect the battery heating rate and effect. The initial temperature of the comparison test was maintained at a room temperature of 25 °C; the battery temperature was not measured, and only the battery current after stabilization was recorded. The motor was not loaded in the test, and it was found that the battery current amplitude was sufficient for comparison under no load, so the motor load torque in the model was also set to zero. The motor used in this test is designed for a low-speed vehicle, and its maximum speed is only 3000 r/min; therefore, three points of 500, 1000, and 1500 r/min were selected in the test, representing creep, low-speed, and medium-speed driving conditions, respectively. In high-speed driving, the output current of the battery is much larger, and its DC heating power is considered to be sufficient to resist the battery temperature drop while driving in a low-temperature environment, so the high-speed driving condition is not the focus of this study.

The simulation and experimental data of the battery current based on SVPWM at a speed of 500 r/min are shown in Figure 10. The numerals 1, 2, 3, and 4 depicted in the this figure and those that in the following text denote the channel designations of the oscilloscope, with the colors of these numerals matching the hues of the corresponding curves in the figure. From a macroscopic point of view, the positive and negative amplitudes of the battery current calculated by the model are basically the same as the experimental values. Microscopically, the experimental value has a harmonic wave of about 4 Hz, resulting in the battery current becoming unstable. The time scale of the harmonic period is much larger than the PWM period and the adjustment period of the PI controller, which is considered to be related to the mechanical characteristics of the motor shaft system. The simulation results also have harmonic components, mainly at a period of 1 ms, which is the same as the control period of the speed loop, so it is speculated that it is caused by the regulation of the speed loop PI controller. Battery current comparison data at speeds of 1000 and 1500 r/min is shown in Figures 11 and 12, respectively. On the whole, the increase in speed will cause the positive battery current to increase proportionally while the battery charge current does not increase significantly.

In order to observe the adjusted effect of the dSVPWM algorithm on battery heating intensity, the dead zone is added to the working conditions of 450 r/min. Comparison results between the dSVPWM algorithm, with $b_n$ = 20% and 50%, and the original SVPWM algorithm, with $b_n$ = 0, are shown in Figure 13. It can be seen that with the increase of $b_n$ in the dSVPWM algorithm, an obvious charge current of the battery emerges, and the discharge current of the battery also increases significantly. In addition, the motor speed was quite stable during the experiment. This shows that the dSVPWM algorithm can superimpose a large AC on the battery's DC discharge current to realize the self-heating of the battery. More importantly, the amplitude of the superimposed AC can be adjusted by $b_n$ to realize the adjustable and controllable heating intensity of the battery.

Through the above experimental data, the theoretical feasibility of the dSVPWM algorithm in motor speed control is preliminarily verified. Through the comparison of experimental and model calculation results, the accuracy of model structure and parameters is verified. The experimental data also prove the controllability of the dSVPWM algorithm on the battery AC. Based on this, the feasibility study on battery pack heating based on the dSVPWM algorithm will be further carried out in the following sections of the simulation.

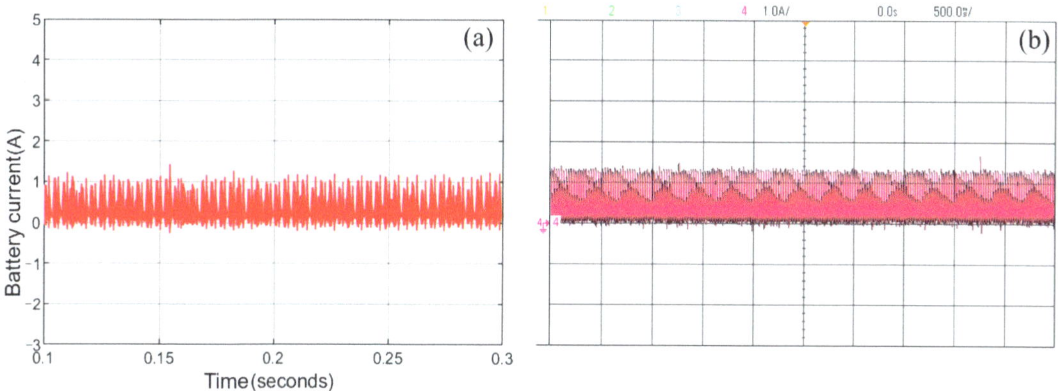

**Figure 10.** Comparison of the battery current calculated by the model and measured in the experiment based on SVPWM at a fixed speed of 500 r/min: (**a**) simulation data, (**b**) experimental data.

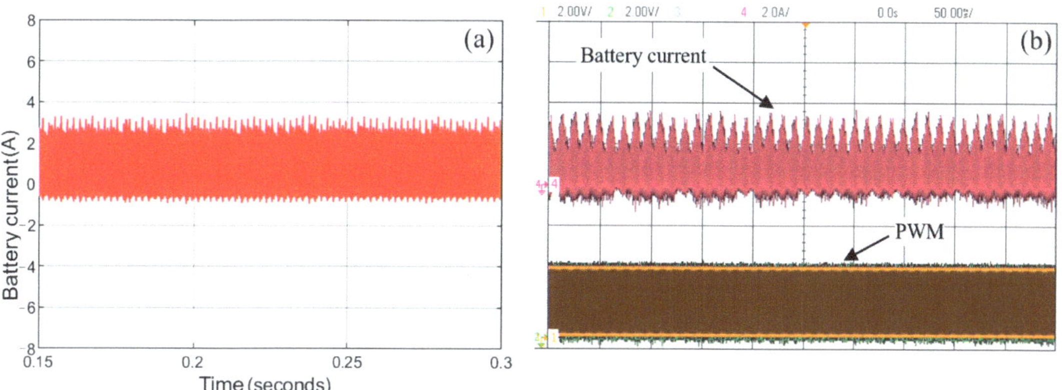



**Figure 11.** Comparison of the battery current calculated by the model and measured in the experiment based on SVPWM at a fixed speed of 1000 r/min: (**a**) simulation data, (**b**) experimental data.

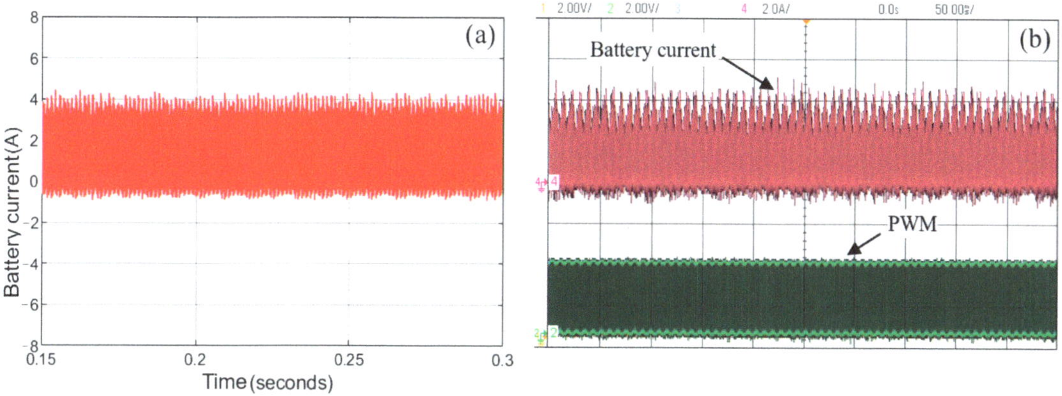



**Figure 12.** Comparison of the battery current calculated by the model and measured in the experiment based on SVPWM at a fixed speed of 1500 r/min: (**a**) simulation data, (**b**) experimental data.

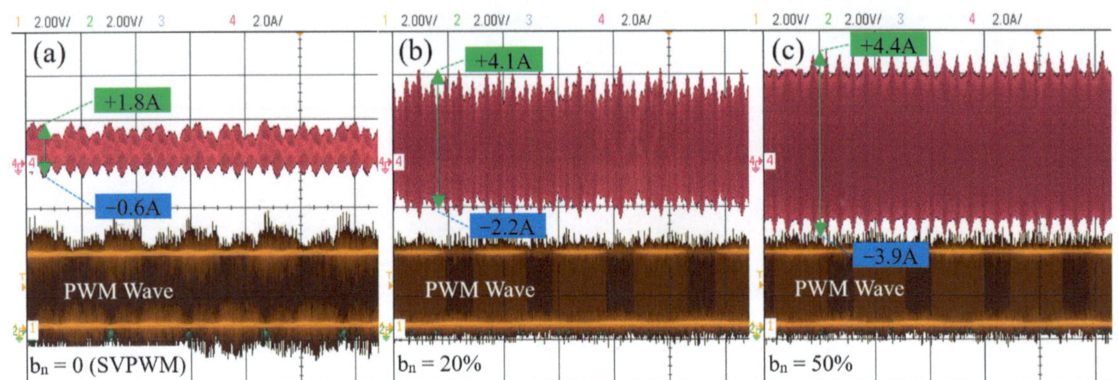



**Figure 13.** The effect of the dSVPWM algorithm on modifying the battery current in the experiment: (**a**) $b_n = 0$, (**b**) $b_n = 20\%$, (**c**) $b_n = 50\%$.

### *4.2. Comparison of dSVPWM and SVPWM Algorithm Using Simulation*

#### 4.2.1. Comparison of Motor Phase Current

The target speed was set to 2500 r/min, the motor load was set to a constant 1.0 Nm, and both $b_n$ and $a_{cx}$ were set to 1. The initial and ambient temperatures of the battery were set to 20 °C. The motor phase currents at 0.3 s for dSVPWM and SVPWM algorithms were recorded, respectively, as shown in Figure 14. At this time, the motor speed had been stabilized. The figure shows that there is no significant difference in the mean value of the phase current, while the peak current of the dSVPWM algorithm increases slightly, and its ripple is larger, especially near the peak. This is due to the commutation of the voltage vector caused by the switching of the non-zero vector and dead zone, and the voltage difference is almost twice that of SVPWM, so a larger ripple current is formed. In general, the sinusoidal fluctuation of the motor phase current can still be maintained in the dSVPWM algorithm, and its root-mean-square value and the phase difference have not changed significantly under the same working conditions.

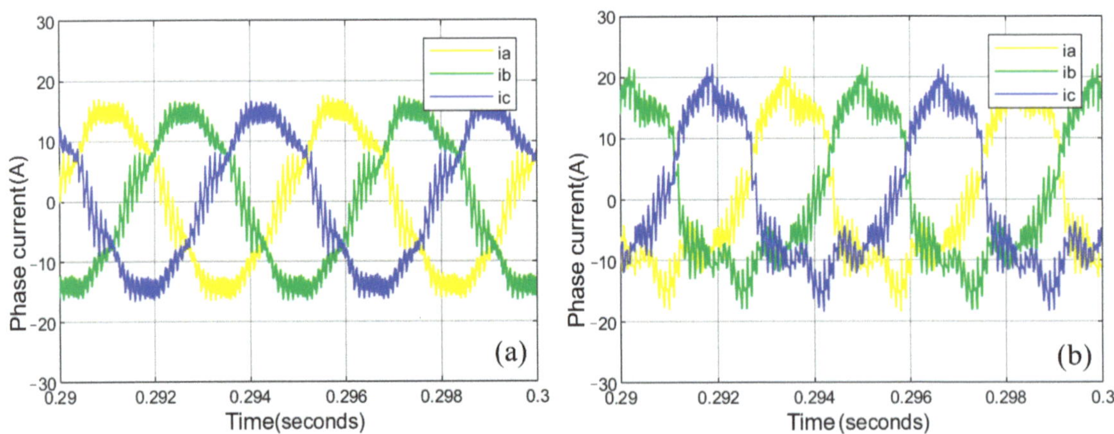



**Figure 14.** Comparison of motor phase current: (**a**) SVPWM algorithm, (**b**) dSVPWM algorithm.

#### 4.2.2. Comparison of Battery Current

The comparison of battery currents under the same operating parameters as Section 4.2.1 is shown in Figure 15. Obviously, the control strategy using the SVPWM algorithm does not form AC battery current, while the motor control strategy using the dSVPWM algorithm generates alternating currents in the battery, and the amplitude of the current is greatly increased. From Figure 15c, the local magnification of Figure 15b, around 0.3 s, it can be seen that the period of AC battery current coincides with the PWM period $T_s$ and that the positive and negative peaks of the alternating current are not equal. The positive peak is greater than the negative peak, and the proportion of the positive and negative values of the alternating current in time is quite different, which is determined by the proportion of the dead zone in the dSVPWM algorithm. Especially in high-speed conditions, non-zero vector accounts for the majority of a PWM period, and the proportion of dead zone is quite small, so the time that negative current accounts for is less than positive current. In addition, similar to the phase current, the peak ripple of the dSVPWM algorithm is significantly higher than that of SVPWM for the same reason as above. Keeping the other operating parameters unchanged, the target speed was set to 500 r/min. This simulation result of the battery current is shown in Figure 15d, which shows that the amplitude of the alternating current at low speed has been greatly improved. This is because, at low speeds, the dSVPWM algorithm has a larger proportion of zero vectors, giving more time to generate AC without changing the $b_n$.

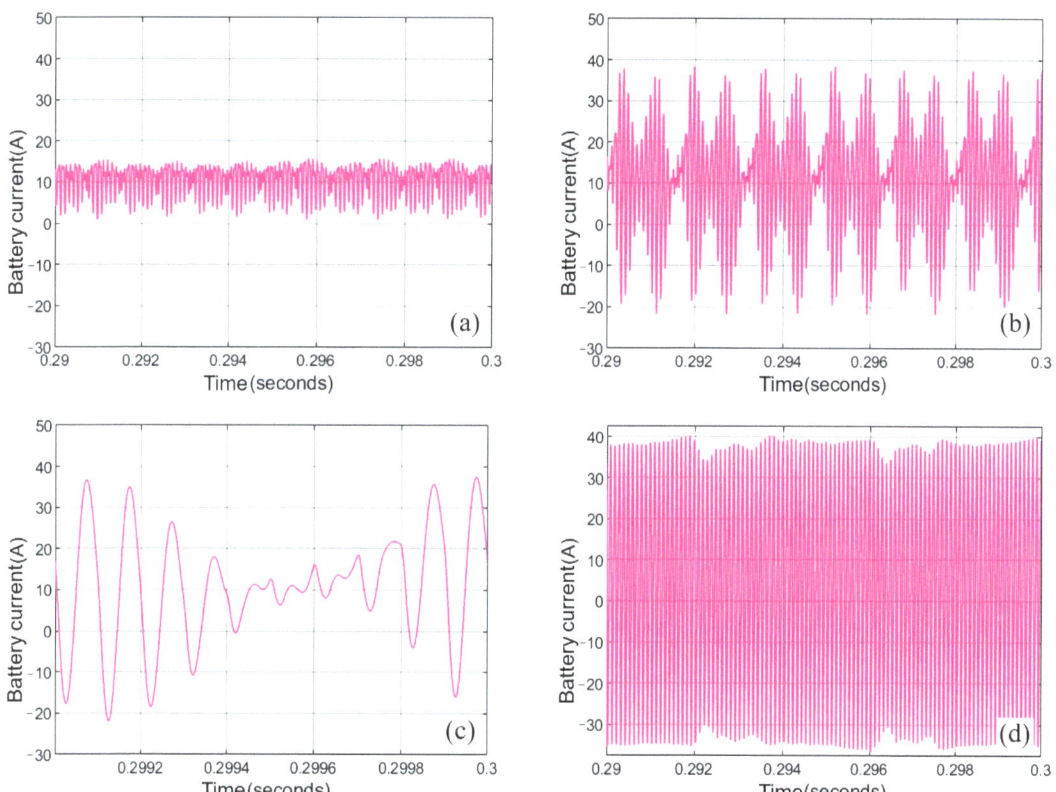

**Figure 15.** Comparison of battery current: (**a**) SVPWM algorithm, (**b**) dSVPWM algorithm at high speed, (**c**) local magnification of (**b**), (**d**) dSVPWM algorithm at low speed.

4.2.3. Comparison of Step Response to Torque Loads at Medium or High Speed

The torque of the motor was set to zero when the time was less than 0.15 s and increased to 1.0 Nm in step from 0.15 to 0.3 s, and the target speed was set to 2500 r/min. Under these conditions and keeping the others the same as in Section 4.2.1, the dynamic process and stability of the two algorithms were compared under the conditions of no load and step load. The results of the simulation are shown in Figure 16, which shows that the two algorithms can remain stable under both no-load and constant-load conditions, and the speed error is zero. Compared with the SVPWM algorithm, the dSVPWM algorithm has a larger overshoot but a smaller rise time. In general, there is not much difference in dynamic performance.

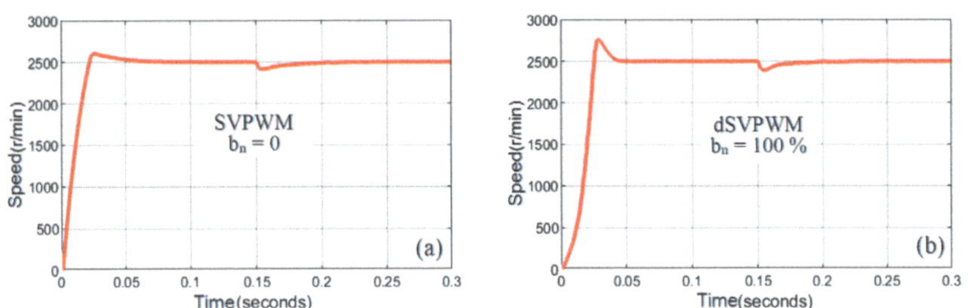

**Figure 16.** Step response to torque loads at medium or high speed: (**a**) SVPWM algorithm, (**b**) dSVPWM algorithm.

### 4.2.4. Stability of Speed Control at Low Speeds and Small Load

The target speed of the motor was set to 500 r/min, and the torque condition remained unchanged. The influence of the two algorithms on the speed control characteristics of the motor under low-speed conditions was studied, and the calculated speed figures are shown in Figure 17. Thereinto, Figure 17a–c indicate that the dSVPWM algorithm can control and maintain a stable speed at low speed and is not affected by the value of $b_n$, but the overshoot is higher than that of the SVPWM algorithm. In addition, compared with Figure 17d, they show that the overshoot of response to the torque disturbance using the dSVPWM algorithm is also larger. But in the end, both algorithms can overcome the torque disturbance and maintain the stability of the speed. In general, the dSVPWM algorithm does not affect the speed control of the motor at both high and low speeds.

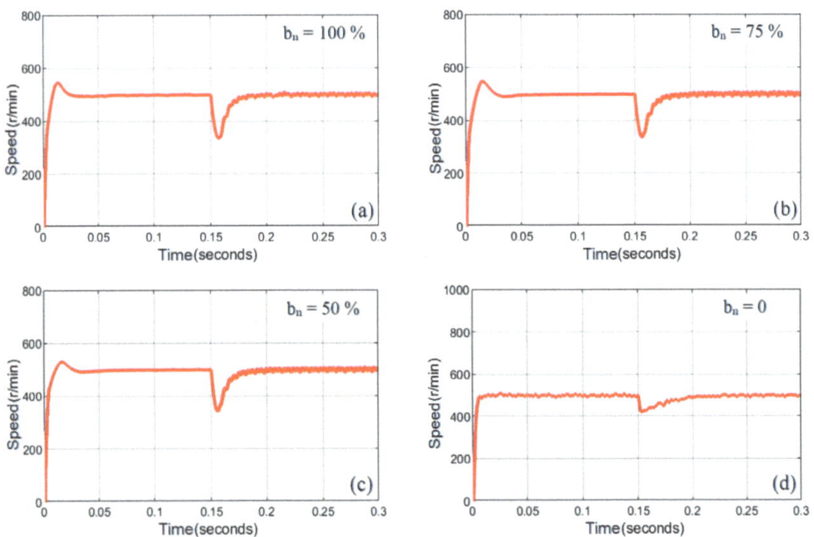

**Figure 17.** Step response to torque loads at low speed: (**a**–**c**) dSVPWM algorithm at different $b_n$, (**d**) SVPWM algorithm.

#### 4.2.5. Comparison of Torque Fluctuations at Low Speed

The target speed of the motor was set to 500 r/min, and the torque condition remained unchanged. Take the simulation result of torque for the last 0.01 s, as shown in Figure 18, which shows that the average torque of the two algorithms is the same, while the torque of the dSVPWM algorithm produces a symmetrical ripple based on the SVPWM algorithm. The frequency of this ripple is the same as the PWM frequency, so it can be inferred that the torque fluctuation is due to the phase current ripple caused by the dead-zone vector. Further comparison of Figure 18a–c shows that the magnitude of the torque ripple decreases with the decrease of $b_n$, which further indicates that the ripple is caused by the dead zone added to the dSVPWM algorithm. Due to the high frequency and small amplitude of this torque ripple, 10 kHz and less than $\pm 1$ Nm, it is easily filtered by mechanical systems with large time constants. In addition, the simulation results show that with the increase in speed, the proportion of dead zone decreases, and the amplitude of the torque ripple will also decrease accordingly.

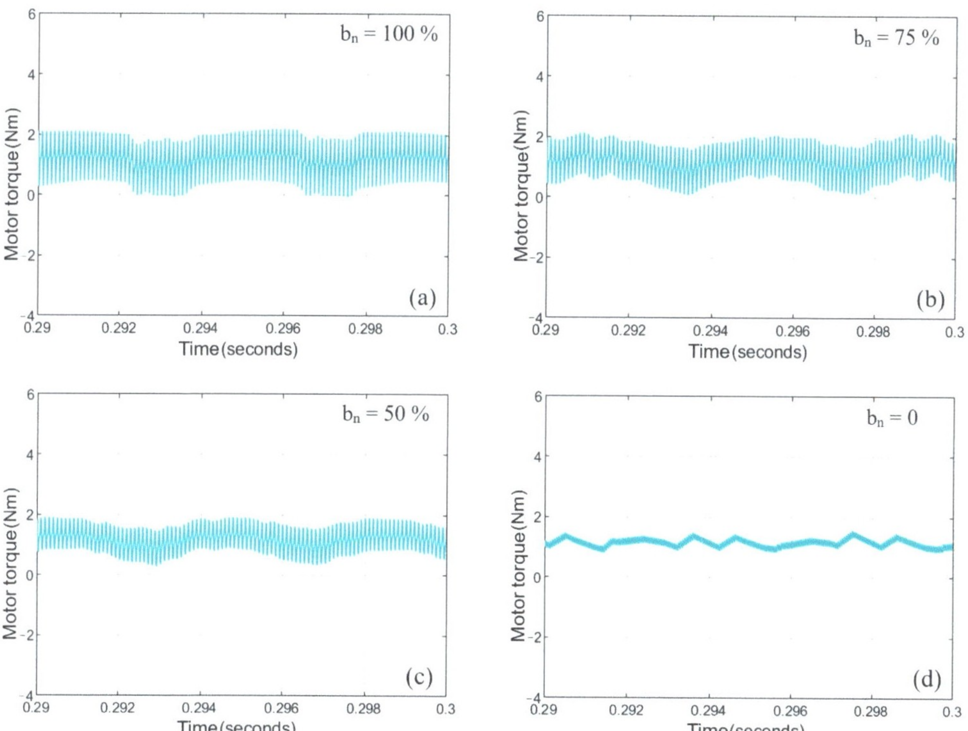


**Figure 18.** Motor torque ripple at low speed: (**a–c**) dSVPWM algorithm at different $b_n$, (**d**) SVPWM algorithm simulation results of temperature-locking strategy under CLTC driving conditions.

#### 4.2.6. Simulation Results Related to Vehicle-Level Powertrain Model

The simulation results are shown in Figure 19, which includes (a) motor speed tracking accuracy; (b) peak and trough value per PWM cycle of battery voltage; (c) peak and trough value per PWM cycle of battery current; (d) heating intensity $b_n$ and the comparison of the average battery temperature with or without temperature-locking strategy. Figure 19a explicitly shows that the motor can track the target speed under CLTC conditions well, and the speed is almost error-free, without overshoot and jitter. As can be seen from Figure 19b, the battery current has a large amplitude during driving, but it is unstable and changes drastically with the load. Especially during the rapid acceleration and deceleration of the vehicle, there are several large spikes. Microscopically, during each PWM period, there is a negative battery current, which is caused by the current fluctuations in the d-axis caused by PI regulation. Macroscopically, the battery current is positive, indicating discharge. When the vehicle is stopped, the battery current is close to zero. When the temperature-locking strategy is in effect, a large magnitude of AC can be immediately generated on the battery DC link. Its peak-to-peak values can reach up to 200 A. More importantly, this AC can be superimposed on the normal operating current without affecting the speed tracking of the motor. As can be seen in Figure 19c, the battery voltage is always maintained between 225 and 378 V. Even in the 50th~70th s, the battery voltage does not exceed the upper and lower limits when the battery not only needs to power the motor but also to heat itself. This is due to the fact that, under the effect of the voltage-limiting control strategy, $b_n$ has been called back to 0 many times so that the voltage falls back to the safe range in time.

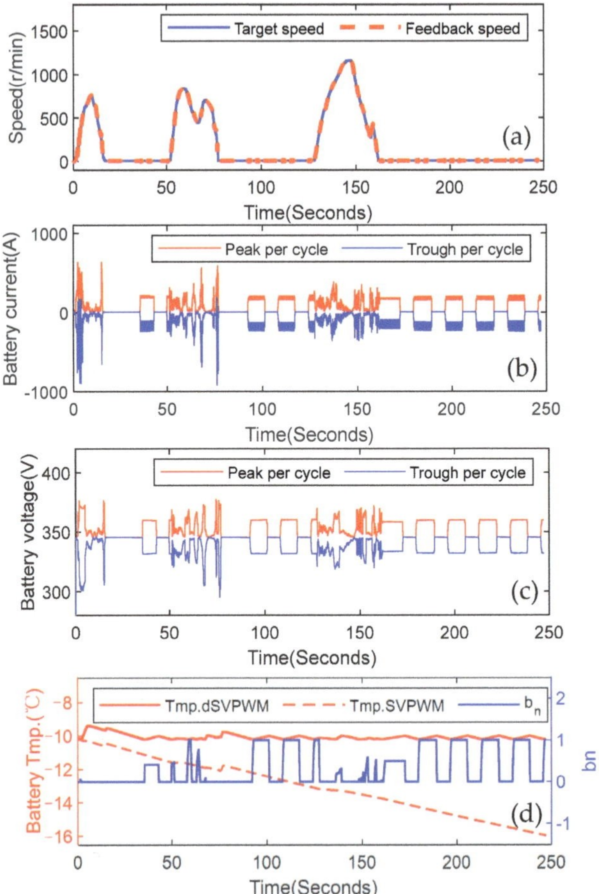

**Figure 19.** Simulation results using a fragment of the CLTC condition: (**a**) motor speed, (**b**) battery current, (**c**) battery voltage, (**d**) battery temperature and $b_n$.

As shown in Figure 19d, the third and fourth pulses of $b_n$ between 50 and 70 s are caused by voltage exceeding the upper limit. This implies that the temperature-locking with voltage-limiting strategy is effective in real driving conditions. The average battery temperature plotted in Figure 19d rises significantly after the start of the first driving—stopping cycle, where the heating intensity is zero. When the vehicle stopped for the first time, battery temperature dropped to $-10.2\ ^\circ$C around the 35th s, triggering the temperature locking strategy, and the $b_n$ rose to 0.4. Then, when the temperature reached $-9.8\ °C$, the $b_n$ dropped back to 0. In addition, it is noteworthy that in the third driving–stopping cycle, although the vehicle was driving, it was not possible to maintain its battery temperature due to the small current, so the temperature locking strategy was repeatedly activated, locking the battery temperature above $-10\ °C$ at all times. Throughout the extended standstill phase, the temperature-locking strategy is repeatedly activated, resulting in the battery temperature fluctuating between $\pm 0.2\ °C$. This maintains the vehicle's readiness for immediate startup. In short, the temperature-locking strategy successfully locked the lowest battery temperature at around $-10\ °C$. In contrast, the temperature of the no-lock strategy, as shown by the red dotted line in Figure 19d, generally decreased from $-10\ °C$ to $-16\ °C$, which was obviously unable to maintain the normal operation of the battery in an extremely cold environment.

### 4.2.7. Calculation and Discussion of Energy Balance and Battery Heating Efficiency

According to the aforementioned analysis, the proposed dSVPWM algorithm does not change the magnitude and direction of the FOC vector at the micro level, and the maximum modulation is consistent with the SVPWM algorithm. At the same time, the proposed method does not alter the structure of the motor's stator and rotor. Therefore, the proposed control strategy does not reduce the efficiency of motor control. Fundamentally, the efficiency of transmuting electrical energy stored in the battery into mechanical energy persists unaltered. Upon examining the aspect of battery heating, the primary concern evolves to the effectiveness of converting electrical energy into the internal energy encapsulated within the battery.

According to the law of energy conservation, the electrical energy stored in the battery gradually decreases throughout the self-heating process. This electrical energy can be divided into two parts: one part is converted into electromagnetic energy, which is used to drive the motor, and the other part is used for heating the battery. The emphasis of this study is on the electrical energy allocated for heating the battery. This segment of electrical energy is partly transformed into the battery's internal energy, which serves to elevate the battery's temperature, whereas some of this internal energy is dissipated into the surrounding environment via heat exchange at the battery's surface. The remaining electrical energy is converted into heat energy via the equivalent resistance of the external circuitry.

Based on this, the article defines the heating efficiency as follows:

$$\eta_{\text{heat}} = \frac{Q_{\text{ine}}}{E_{\text{ele}}} = \frac{Q_{\text{ine}}}{Q_{\text{ine}} + Q_{\text{dis}} + Q_{\text{cct}}} \tag{16}$$

where in $Q_{\text{ine}}$ represents the increment of internal energy of the battery, $E_{\text{ele}}$ represents the diminishment of the battery's electrical energy attributable to the heating process, $Q_{\text{dis}}$ represents the heat dissipation of the battery, and $Q_{\text{cct}}$ represents the heat production from the external circuit.

$E_{\text{ele}}$ can be calculated according to the battery open-circuit voltage-capacity curve, as shown in the following equation:

$$E_{\text{ele}} = \int_{Q_0}^{Q_{\text{eoh}}} u_{\text{ocv}} dQ \tag{17}$$

Among them, $Q_0$ represents the battery capacity at the start of heating, $Q_{\text{eoh}}$ represents the battery capacity at the end of heating, and $u_{\text{OCV}}$ represents the battery open-circuit voltage corresponding to different capacities. Therefore, as long as the SOC of the battery before and after heating is known, the battery's energy consumption can be obtained.

$Q_{\text{ine}}$ can be calculated according to the following equation:

$$Q_{\text{ine}} = c_b m_b (T_{\text{eoh}} - T_0) \tag{18}$$

wherein $c_b$ represents the total specific heat capacity of the battery, $m_b$ represents the mass of the battery, $T_0$ represents the initial temperature of the battery, and $T_{eoh}$ represents the temperature at the end of battery heating.

$Q_{dis}$ can be calculated according to the following equation:

$$Q_{dis} = \int_0^{t_{eoh}} hA(T_b - T_{amb})dt \tag{19}$$

Here, $h$ represents the surface heat transfer coefficient, $A$ represents the surface area of the battery, $T_b$ represents the average temperature of the battery surface at time $t$, $T_{amb}$ represents the ambient temperature, and $t_{eoh}$ represents the time used for heating.

Assuming in a PWM cycle, a non-zero vector acts for $t_{act}$ seconds, followed by the insertion of a dead-zone vector for $t_{dead}$ seconds. The calculation equation for $Q_{cct}$ is as follows:

$$Q_{cct} = \frac{3}{2}\int_0^{t_{act}} i_{dc}^2(R_{on} + R_s)dt + \frac{3}{2}\int_{t_{act}}^{t_{act}+t_{dead}} i_{dc}^2 R_s dt + 2\int_0^{t_{act}+t_{dead}} i_b^2 R_{pl} dt \tag{20}$$

In this equation, $i_{dc}$ represents the DC-link current, $i_b$ represents the battery current, $R_{on}$ represents the on-resistance of the MOSFET, $R_s$ represents the phase resistance of the motor, and $R_{pl}$ represents the equivalent ohmic resistance of a single DC-link. The calculation of $Q_{cct}$ above does not take into account the parasitic resistance of the DC-link capacitor because its magnitude is small.

Based on the aforementioned equation, it is evident that the heating efficiency is related to the degree of surface insulation of the battery and the real part of the impedance of the external circuit of the battery. Assuming the surface insulation of the battery is good, the real part of the external circuit impedance becomes the key indicator for evaluating the heating efficiency. The ohmic impedance of a single cell between $-30$ and $-10\,^\circ$C is approximately between 20 and 35 m$\Omega$, which is more than twice its normal temperature impedance. Since the series number of the battery pack is often much greater than the parallel number, the total impedance of the battery pack increases by an order of magnitude on the basis of the cell impedance. For example, a 380 V battery pack requires 102 series of NCM cells. Assuming the parallel number is 3, its total ohmic internal resistance would be between 680 and 1190 m$\Omega$. In comparison, the equivalent resistance of the external circuit is limited in both value and quantity and is not highly sensitive to temperature. For example, the on-state resistance of an efficient MOSFET is about 3 m$\Omega$, the phase resistance of the motor is about 15 m$\Omega$, and the equivalent resistance of the DC-link is also around 10 m$\Omega$. Therefore, the reduction of electrical energy in the battery during heating is mostly applied to the battery's ohmic internal resistance to increase the battery's own temperature. Based on the above example data, under the condition of battery insulation, the heating efficiency will reach between 95.8% and 97.6%. The primary determinant of heating efficiency is the degree of insulation condition present on the battery's surface.

From the above analysis, it is clear that the majority of the electrical energy reduction during the battery heating process is used for self-heating. The increase in the battery's internal energy can be calculated using Equation (18), and by combining this with the heating efficiency derived from the analysis, the reduction in the battery's State of Charge (SOC) during the heating process can be estimated. Taking the battery sample studied in Section 2.4 as an example, at environmental temperature and the initial battery temperature of $-30\,^\circ$C, heating up to $-10\,^\circ$C results in an internal energy increase of approximately 0.25 Wh per cell. Assuming a heating efficiency of 96%, the total energy consumption for the heating process is about 0.26 Wh. If the open-circuit voltage during the heating process is assumed to be around the rated value of 3.7 V, the equivalent capacity consumption would be approximately 70.27 mAh, which translates to a SOC decrease of 2.81%. Similarly, if the temperature is increased from $-30\,^\circ$C to $0\,^\circ$C, the SOC decrease can be calculated to be 4.22%. It is worth noting that the heating efficiency estimated by the above method does not account for the iron losses in the motor because it is much smaller than the loss of copper.

To conduct a more detailed quantitative analysis of the energy distribution relationship for the proposed method, a code was developed to compute the battery thermal power and motor output power within the vehicle-level simulation model detailed in Section 3.2 while simultaneously tracking the SOC consumption of the battery. Utilizing the same configuration and driving conditions, the proposed battery temperature-locking method was implemented. The computational outcomes indicate that the vehicle's total range, based on speed integration, is 189.6 m. The SOC of the battery diminished by 0.065%, starting from 75%. By integrating the battery's open circuit voltage curve at −10 °C, that equates to an energy loss of 22.9 Wh. Dividing this by the distance traveled yields a power consumption of 120.8 Wh/km. Concurrently, the total thermal energy produced by the battery amounts to 2 Wh, encompassing both the energy utilized for battery heating and the heat dissipated to the surroundings. It is equivalent to an energy consumption of 10.5 Wh/km, constituting 8.7% of the overall energy consumption. The motor's mechanical energy output is 19.5 Wh, dedicated to propelling the vehicle, with its energy consumption at 102.8 Wh/km, representing 85.1% of the total energy consumption. The residual energy consumption is primarily attributed to the parasitic resistance of the inverter, motor coil, and DC-link. It is worth noting that this simulation does not account for the energy consumption of other electrical or mechanical loads on the vehicle and is representative solely of the −40 °C low-temperature environment and low-speed driving conditions within the CLTC cycle.

## 5. Conclusions

This paper focuses on the problem that the battery temperature cannot be maintained when the electric vehicle is running in extremely cold environments and low-speed conditions. The dead zone is introduced into the SVPWM algorithm for driving PMSM to form the newly proposed dSVPWM algorithm, which is used to reconstruct the dual-PI control model with an FOC control strategy and theoretically opens up a new technical path to realize the battery AC self-heating while driving based on the topology of the existing electric drive system. With the help of voltage vector synthesis, it is proved that the replacement of part of the zero vector in the original SVPWM algorithm with the dead-zone vector and the compensation vector in the dSVPWM algorithm does not change the original output voltage vector, and the amplitude of the battery AC can be changed by adjusting the heating intensity defined by the ratio of the dead zone and the compensation vector to the original zero vector. With the help of the Simulink model and the small-scale motor driving system test bench, the simulation and experimental comparison of the motor control algorithm based on dSVPWM was carried out. Firstly, it is found that the battery current amplitude calculated by the model was consistent with the experimental value under three target speeds using the original SVPWM algorithm, which verifies the precision of the numerical model and PI parameters. Secondly, the heating intensity was changed from 0 to 0.5 in the constant speed test. The comparison between the two figures indicated that an obvious AC battery current was formed, with its amplitude increasing from 1.8 A to 4.4 A, and the motor speed could be kept stable, which verified the controllability of the dSVPWM algorithm on the battery AC. Then, the dSVPWM and SVPWM algorithms were compared and studied in the numerical model, and it is concluded that the dSVPWM algorithm can obtain up to double the peak-to-peak alternating battery current at medium and high speeds, and although the dSVPWM algorithm produces additional motor phase current and torque ripple positively correlated with heating intensity under the same working conditions, it retains excellent resistance to load torque disturbances across a range of target speeds and varying levels of heating intensity, which proves that the proposed algorithm possesses adequate stability. Finally, a structure and workflow for the temperature-locking strategy are given, and the goal is to lock the battery temperature within the set thresholds and avoid the battery voltage overrun. Combined with the parameters of a passenger car provided by the vehicle manufacturer, the proposed temperature locking strategy was deployed into the numerical model of the motor driving system. The study investigated the control effect of the temperature locking strategy on the voltage, current, and temperature of the vehicle power battery under CLTC urban conditions, comparing the battery temperature with that observed under a strategy without temperature locking. It is concluded that (1) the temperature-locking strategy effectively prevents the average battery temperature from declining during urban low-speed driving conditions. Without this strategy, the average battery temperature decreases by 1.44 °C per minute. (2) the strategy ensures precise speed tracking in urban low-speed driving conditions while maintaining the battery temperature within specified upper and lower thresholds. The average temperature fluctuation remains within 1 °C. (3) with the heating function activated, the strategy effectively prevents battery voltage from exceeding safe limits, ensuring the battery's safe operation. In conclusion, the above theories and methods proposed in this paper open up a new feasible technical route for the battery AC self-heating during vehicle driving, and based on this, how to realize the battery AC self-heating coupling with other senior motor control strategies can be further studied.

This paper merely expounds on the fundamental principles of the proposed methodology, accompanied by an initial validation of its feasibility through preliminary examination. Future research endeavors can be pursued in the following aspects: Firstly, incorporate temperature-sensitive parameters of the motor and controller into the simulation model, thereby constructing a heightened comprehensive temperature-adaptive simulation and calculation platform. This platform will facilitate an in-depth examination of the performance characteristics and temperature variations exhibited by the motor and inverter within extremely cold environments. Secondly, embark on a spectral analysis of current and torque data, aiming to meticulously analyze the impact of this methodology on motor torque fluctuations and NVH phenomena, adopting a frequency domain approach for a more profound understanding. Furthermore, from a rigorous energy balance perspective, a quantitative analysis will be conducted to elucidate the relationship between motor drive efficiency and heating efficiency, along with a thorough examination of other intricately related issues.

**Author Contributions:** Conceptualization, W.L.; methodology, W.L.; validation, W.L. and W.S.; investigation, W.L.; data curation, W.L.; writing—original draft preparation, W.L.; writing—review and editing, W.L.; visualization, W.L.; software, W.L. and W.S.; formal analysis, W.S.; resources, S.X.; supervision, S.X.; administration, S.X.; funding acquisition S.X.; project, S.X. All authors have read and agreed to the published version of the manuscript.

**Funding:** This research was funded by the Open Foundation for Key Laboratory of Clean Energy and Carbon Neutrality of Zhejiang Province grant number [204022023005A] and Zhejiang Province Spearhead and Leading Goose Research and Development Key Program grant number [2023C01239].

**Data Availability Statement:** The original contributions presented in the study are included in the article, further inquiries can be directed to the corresponding author.

**Conflicts of Interest:** The authors declare no conflict of interest.

## References

1. Lee, D.Y.; Cho, C.W.; Won, J.P.; Park, Y.C.; Lee, M.Y. Performance characteristics of mobile heat pump for a large passenger electric vehicle. *Appl. Therm. Eng.* **2013**, *50*, 660–669. [CrossRef]
2. Qin, F.; Xue, Q.; Albarracin Velez, G.M.; Zhang, G.; Zou, H.; Tian, C. Experimental investigation on the heating performance of heat pump for electric vehicles at −20 °C ambient temperature. *Energy Convers. Manag.* **2015**, *102*, 39–49. [CrossRef]
3. Zhu, J.; Sun, Z.; Wei, X.; Dai, H.; Gu, W. Experimental investigations of an AC pulse heating method for vehicular high power lithium-ion batteries at subzero temperatures. *J. Power Sources* **2017**, *367*, 145–157. [CrossRef]
4. Jaguemont, J.; Boulon, L.; Dubé, Y. A comprehensive review of lithium-ion batteries used in hybrid and electric vehicles at cold temperatures. *Appl. Energy* **2016**, *164*, 99–114. [CrossRef]
5. Zhang, J.; Ge, H.; Li, Z.; Ding, Z. Internal heaating of lithium-ion batteries using alternating current based on the heat generation model in frequency domain. *J. Power Sources* **2015**, *273*, 1030–1037. [CrossRef]

6.    Zhu, J.; Sun, Z.; Wei, X.; Dai, H. An alternating current heating method for lithium-ion batteries from subzero temperatures. *Int. J. Energy Res.* **2016**, *40*, 1869–1883. [CrossRef]
7.    Ruan, H.; Jiang, J.; Sun, B.; Zhang, W.; Gao, W.; Wang, L.Y.; Ma, Z. A rapid low temperature internal heating strategy with optimal frequency based on constant polarization voltage for lithium-ion batteries. *Appl. Energy* **2016**, *177*, 771–782. [CrossRef]
8.    Li, J.; Sun, D. Lithium-ion Batteries Modeling and Optimization Strategies for Sinusoidal Alternating Current Heating at Low Temperature. *Energy Procedia* **2018**, *152*, 562–567. [CrossRef]
9.    Li, J.-Q.; Fang, L.; Shi, W.; Jin, X. Layered thermal model with sinusoidal alternate current for cylindrical lithium-ion battery at low temperature. *Energy* **2018**, *148*, 247–257. [CrossRef]
10.   Jiang, J.; Ruan, H.; Sun, B.; Wang, L.; Gao, W.; Zhang, W. A low-temperature internal heating strategy without lifetime reduction for large-size automotive lithium-ion battery pack. *Appl. Energy* **2018**, *230*, 257–266. [CrossRef]
11.   Guo, S.; Xiong, R.; Wang, K.; Sun, F. A novel echelon internal heating strategy of cold batteries for all-climate electric vehicles application. *Appl. Energy* **2018**, *219*, 256–263. [CrossRef]
12.   Guo, S.; Xiong, R.; Sun, F.; Cao, J.; Wang, K. An echelon internal heating strategy for lithium-ion battery. *Energy Procedia* **2017**, *142*, 3135–3140. [CrossRef]
13.   Ruan, H.; Jiang, J.; Sun, B.; Wu, N.; Shi, W.; Zhang, Y. Stepwise segmented charging technique for lithium-ion battery to induce thermal management by low-temperature internal heating. In Proceedings of the 2014 IEEE Conference and Expo Transportation Electrification Asia-Pacific (ITEC Asia-Pacific), Beijing, China, 31 August–3 September 2014; IEEE: New York, NY, USA, 2014.
14.   Guo, S.; Xiong, R.; Shen, W.; Sun, F. Aging investigation of an echelon internal heating method on a three-electrode lithium-ion cell at low temperatures. *J. Energy Storage* **2019**, *25*, 100878. [CrossRef]
15.   Shang, Y.; Xia, B.; Cui, N.; Zhang, C.; Mi, C.C. An Automotive Onboard AC Heater Without External Power Supplies for Lithium-Ion Batteries at Low Temperatures. *IEEE Trans. Power Electron.* **2018**, *33*, 7759–7769. [CrossRef]
16.   Shang, Y.; Zhu, C.; Fu, Y.; Mi, C.C. An Integrated Heater Equalizer for Lithium-Ion Batteries of Electric Vehicles. *IEEE Trans. Ind. Electron.* **2019**, *66*, 4398–4405. [CrossRef]
17.   Shang, Y.; Liu, K.; Cui, N.; Zhang, Q.; Zhang, C. A Sine-Wave Heating Circuit for Automotive Battery Self-Heating at Subzero Temperatures. *IEEE Trans. Ind. Inform.* **2020**, *16*, 3355–3365. [CrossRef]
18.   Shang, Y.; Liu, K.; Cui, N.; Wang, N.; Li, K.; Zhang, C. A Compact Resonant Switched-Capacitor Heater for Lithium-Ion Battery Self-Heating at Low Temperatures. *IEEE Trans. Power Electron.* **2020**, *35*, 7134–7144. [CrossRef]
19.   Zhang, Y.; Yang, Y.; Shang, Y.; Cui, N. A high frequency AC heater based on switched capacitors for lithium-ion batteries at low temperature. *J. Energy Storage* **2021**, *42*, 102977. [CrossRef]
20.   Zhu, C.; Zhang, H.; Lu, F. A Compact Onboard Battery Self-Heater for All-Electric Aircraft Applications at Cold Climates. In Proceedings of the 2019 IEEE National Aerospace and Electronics Conference (NAECON), Dayton, OH, USA, 15–19 July 2019; IEEE: New York, NY, USA, 2019.
21.   Zhu, C.; Shang, Y.; Lu, F.; Zhang, H. Optimized Design of an Onboard Resonant Self-Heater for Automotive Lithium-Ion Batteries at Cold Climates. In Proceedings of the 2019 IEEE Energy Conversion Congress and Exposition (ECCE), Baltimore, MD, USA, 29 September–3 October 2019; IEEE: New York, NY, USA, 2019.
22.   Zhu, C.; Cao, Y.; Zhang, H.; Lu, F.; Zhang, X. Comprehensive Design and Optimization of an Onboard Resonant Self-Heater for EV Battery. *IEEE Trans. Transp. Electrif.* **2021**, *7*, 452–463. [CrossRef]
23.   Baba, H.; Kawasaki, K.; Kawachi, H. *Battery Heating System for Electric Vehicles*; SAE Technical Paper Series; SAE: Warrendale, PA, USA, 2015.
24.   Li, Y.; Gao, X.; Qin, Y.; Du, J.; Guo, D.; Feng, X.; Lu, L.; Han, X.; Ouyang, M. Drive circuitry of an electric vehicle enabling rapid heating of the battery pack at low temperatures. *iScience* **2021**, *24*, 101921. [CrossRef]
25.   Li, Y.; Du, J.; Zhou, G.; Ouyang, M.; Fan, Y. A rapid self-heating battery pack achieved by novel driving circuits of electric vehicle. *Energy Rep.* **2020**, *6*, 1016–1023. [CrossRef]
26.   Zhu, C.; Han, J.; Guo, B.; Fan, G.; Zhang, X. A Split-Source Self-Heater for Automotive Batteries Based on Traction Drive Reconfiguration. *IEEE J. Emerg. Sel. Top. Ind. Electron.* **2023**, *4*, 188–197. [CrossRef]
27.   Du, C.; Peng, Q.; Chen, F.; Deng, K.; Chen, J.; Deng, C.; Hu, M. Investigation on the method of battery self-heating using motor pulse current. *Proc. Inst. Mech. Eng. Part D J. Automob. Eng.* **2022**, *236*, 2399–2409. [CrossRef]
28.   Li, W.; Shi, W.; Xiong, S.; Huang, H. Voltage-feedback Internal Heating Strategy of Lithium-Ion Battery in an Extremely Cold Environment Implemented with Drive System of Electric Vehicle. *IEEE Trans. Transp. Electrif.* **2023**. [CrossRef]
29.   Yuan, L.; Hu, B.; Wei, K.; Chen, Z. *Modern Permanent Magnet Synchronous Motor Control Principle and MATLAB Simulation*; Beihang University Press: Beijing, China, 2016.
30.   Harnefors, L.; Pietilainen, K.; Gertmar, L. Torque-maximizing field-weakening control, design, analysis, and parameter selection. *IEEE Trans. Ind. Electron.* **2001**, *48*, 161–168. [CrossRef]

*World Electric Vehicle Journal*

MDPI

*Article*

# Online Fault Detection of Open-Circuit Faults in a DTP-PMSM Using Double DQ Current Prediction

**Qiang Geng [1], Wenhao Du [1], Xuefeng Jin [1], Guozheng Zhang [1] and Zhanqing Zhou [2,***

[1] Tianjin Key Laboratory of Intelligent Control of Electrical Equipment, Tiangong University, Tianjin 300387, China; gengqiang@tju.edu.cn (Q.G.); duwenhao@tiangong.edu.cn (W.D.); jinxuefeng@tiangong.edu.cn (X.J.); zhanggz@tju.edu.cn (G.Z.)

[2] Advanced Electrical Equipment Innovation Center, Zhejiang University, Hangzhou 311107, China

* Correspondence: zhouzhanqing@tiangong.edu.cn

**Abstract:** This research proposes a strategy to diagnose open-phase faults (OPF) and open-switching faults (OSF) in dual three-phase permanent magnet synchronous motor (DTP-PMSM) inverters. The method is based on the dual d–q predictive current model and involves establishing a mathematical model and utilizing the finite control set model predictive current extraction technique to predict the motor current. It then analyzes the characteristics of the switching-tube current under both normal and fault conditions. Finally, a fault predictive current model is introduced and the residual is calculated based on the predicted fault current value and the actual measured current value to diagnose the inverter fault. The proposed method effectively overcomes misjudgment issues encountered in traditional open-circuit fault diagnosis of inverters. It enhances the system's response speed during dynamic processes and strengthens the robustness of diagnosis algorithm parameters. The experimental results demonstrate that the proposed method can rapidly, effectively, and accurately diagnose open-circuit faults presented in this paper fastest within one-fifth of a current cycle. It achieves a diagnostic accuracy rate of 97% in the dual three-phase permanent magnet synchronous motor drive system.

**Keywords:** dual three-phase permanent magnet synchronous motor; model current prediction; inverter fault diagnosis; open-circuit fault

**Citation:** Geng, Q.; Du, W.; Jin, X.; Zhang, G.; Zhou, Z. Online Fault Detection of Open-Circuit Faults in a DTP-PMSM Using Double DQ Current Prediction. *World Electr. Veh. J.* **2024**, *15*, 204. https://doi.org/10.3390/wevj15050204

Academic Editor: Joeri Van Mierlo

Received: 16 April 2024
Revised: 6 May 2024
Accepted: 7 May 2024
Published: 8 May 2024

## 1. Introduction

The development of AC transmission has led to increasing research on high-power control systems. One area of interest for scholars is multiphase motors, which are characterized by low voltage, high power, small torque ripple, and strong fault tolerance [1–3]. Dual three-phase permanent magnet synchronous motors have been extensively studied among multiphase motors. These motors have several advantages over ordinary three-phase motors, including reduced torque ripple and improved fault tolerance. As a result, dual three-phase permanent magnet synchronous motors are an excellent choice for applications requiring high power and high reliability, meeting the demanding performance requirements of the current engineering field [4,5].

The power switching devices in motor drive systems play a crucial role, but they often experience frequent switching states over long periods o, leading to switching losses and heat build-up. In addition, these devices are susceptible to environmental factors, transient conditions during operation, and overload situations that increase the risk of failure. Short-circuit faults are particularly serious as they can quickly propagate throughout the system [6]. Currently, hardware circuits such as fast fuses are used to protect power-switching devices from short-circuit faults by converting them to open-circuit faults once they occur. This helps to prevent further damage to the system. However, unlike short-circuit faults, which immediately stop the operation of the system, open-circuit faults do not necessarily lead to an immediate shutdown but can affect the performance of the motor. If open-circuit faults are not recognized and rectified immediately, they can lead to secondary faults in the system and cause major damage [7]. Therefore, the effective diagnosis of open-circuit faults in inverter switchgear is of great importance to ensure the reliability and safety of the system.

The core of fault detection and localization technology lies in extracting key signal features and comparing them with benchmark parameters to evaluate deviations during equipment operation [8], providing data support for fault diagnosis and localization. Open-circuit faults can significantly affect the system's output current and voltage in inverter applications. Current signals, as essential monitoring indicators in motor control systems, are often selected for open-circuit fault diagnosis using methods such as phase current average value [9], radius [10], angle [11], Park vector [12], zero crossing (ZC) feature [13,14], and similarity feature [15].

To simplify fault detection, a method based on the unique x–y plane current of multiphase motors was proposed in reference [16]. This method utilizes VSD to detect open-phase faults in six-phase induction motors. Similarly, reference [17] proposes a fault diagnosis method for a single switch open-circuit fault in a six-phase fault-tolerant permanent magnet synchronous motor system. This method is based on the average current Park's vector and compares the modulus of the average current Park's vector before and after the switch fault in two orthogonal subspaces to determine whether a switch fault has occurred. On the other hand, voltage-based open-circuit fault diagnosis methods typically require additional voltage sensors in the system. These methods diagnose faults by measuring the phase voltage, line voltage, or voltage values at key points output by the inverter, as discussed in references [18–22]. After considering factors such as neutral point voltage imbalance and time-offset injection, references [23] proposed an assumption-based method for T-type three-level inverter OC fault diagnosis based on the output phase voltage model. The method established a phase voltage model and determined the fault location based on the amplitude and angle of the residual vector. The proposed method in [24] presents a technique for rapid open-circuit fault diagnosis on an induction motor drive system by calculating the common-mode voltage, enabling diagnosis within five control cycles. However, this approach necessitates the design of a complex observer to estimate the motor back-EMF.

In summary, due to their similar characteristics, previous fault diagnosis methods have faced difficulties in achieving simultaneous rapid detection of OPF and OSF. Therefore, this article proposes an online fault diagnosis scheme based on dual dq current prediction by combining model-based and signal-based diagnosis. The proposed strategy can be divided into three steps: first, detecting differences between normal and faulty models of permanent magnet synchronous motors to determine when faults occur; secondly, using different OPF models to predict general deviations between current and dq axis feedback currents for identifying faulty phases; finally, utilizing harmonic currents' trajectory angles for specific fault diagnosis and localization. By employing this method, quick diagnoses of OSF and OPF in PMSM drivers can be achieved within one-fifth of a current cycle.

First, the topological structure and the basic principle of the DTP-PMSM are presented and a mathematical model of the DTP-PMSM is derived. Subsequently, the characteristics of the current at open fault are analyzed and the basic principle and methodology of the fault current model are explained in more detail. Furthermore, the use of the fault current model for the diagnosis of open faults in PMSMs is presented. Finally, the effectiveness of the proposed methodology is validated by experiments, and its feasibility in practice is discussed.

## 2. Mathematical Model of Dual Three-Phase PMSM

The main circuit topology of the dual three-phase PMSM driving system is shown in Figure 1. The stator consists of two groups of three-phase, symmetrical windings that are connected in a Y-shape. The two groups of windings are spatially separated by an electrical angle of 30° and are fed by a six-phase voltage source inverter. To simplify research and analysis, it is assumed that the back EMF of the motor is sinusoidally distributed.

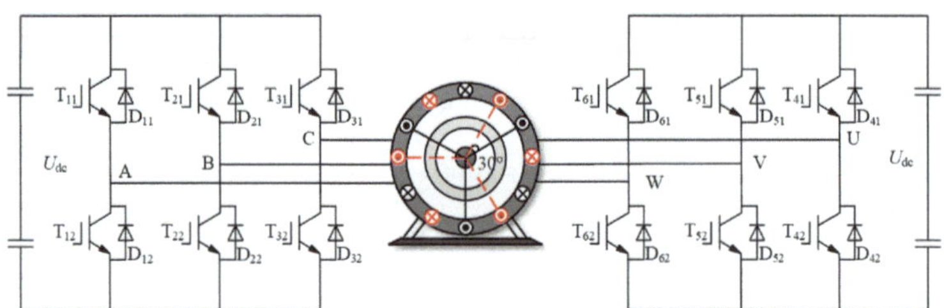

**Figure 1.** Dual three-phase PMSM drive system diagram.

The expression for the stator voltage of a dual three-phase PMSM is

$$U_{6s} = R_{6s}I_{6s} + \frac{d\psi_{6s}}{dt} \tag{1}$$

The $U_{6s}$, $I_{6s}$, $\psi_{6s}$, and $R_{6s}$, respectively, denote the stator voltage vector, phase current vector, magnetic flux vector, and stator resistance.

The vector space decoupling of the dual three-phase PMSM model in the d–q, x–y, o1–o2 coordinate systems for the continuous-time-domain voltage state equation

$$\begin{bmatrix} u_d \\ u_q \\ u_x \\ u_y \end{bmatrix} = R\begin{bmatrix} i_d \\ i_q \\ i_x \\ i_y \end{bmatrix} + \begin{bmatrix} L_d & 0 & 0 & 0 \\ 0 & L_q & 0 & 0 \\ 0 & 0 & L_z & 0 \\ 0 & 0 & 0 & L_z \end{bmatrix}\frac{d}{dt}\begin{bmatrix} i_d \\ i_q \\ i_x \\ i_y \end{bmatrix} + \omega\begin{bmatrix} -L_q i_d \\ L_d i_q + \varphi_f \\ 0 \\ 0 \end{bmatrix} \tag{2}$$

In the equation, $L_d$ and $L_q$ represent the dq-axis inductances, respectively; $L_Z$ denotes the stator side leakage inductance; $i_d$ and $i_q$ are the dq-axis stator currents; $i_x$ and $i_y$ are the current components in the harmonic plane; $i_{o1}$ and $i_{o2}$ are the zero-sequence sub-plane currents; $R_s$ is the stator resistance; $\omega$ is the angular frequency, and $\psi_f$ is the magnitude of the magnetic flux linkage.

The widely used forward Euler method is chosen to discretize the current in Equation (2), as the influence of zero sequence subspace can be neglected due to center point isolation. The predicted values of current in each subspace are obtained accordingly.

$$\begin{cases} i_d^{(k+1)} = (1 - \frac{T_s R_s}{L_d})i_d^{(k)} + \frac{T_s}{L_d}u_d^{(k)} + T_s\omega i_q^{(k)} \\ i_q^{(k+1)} = (1 - \frac{T_s R_s}{L_q})i_q^{(k)} + \frac{T_s}{L_q}u_q^{(k)} - T_s\omega i_d^{(k)} - \frac{T_s\omega\psi_f}{L_q} \\ i_x^{(k+1)} = (1 - \frac{T_s R_s}{L_{ls}})i_x^{(k)} + \frac{T_s}{L_{ls}}u_x^{(k)} \\ i_y^{(k+1)} = (1 - \frac{T_s R_s}{L_{ls}})i_y^{(k)} + \frac{T_s}{L_{ls}}u_y^{(k)} \end{cases} \tag{3}$$

## 3. Fault Analysis

Figure 2 illustrates the voltage and current distortion caused by inverter bridge arm faults, using the W-phase current as an example.

Following the $T_{61}$ switch fault, the positive W-phase current is forced to zero within several switching cycles, as shown in Figure 2. The duration of these few cycles is negligible when compared to a full period. Consequently, the $T_{61_O}$SF fault exhibits similar fault characteristics to the W-phase OPF in the positive half cycle, and the PMSM model with W-phase OPF applies to the $T_{61_O}$SF. During the negative half-cycle of the W-phase current, it does not pass through $T_{61}$; hence, the $T_{61_O}$SF does not affect the W-phase current in the negative half-cycle. Analysis results can be obtained when an open-circuit fault occurs at

$T_{61}$. In addition, as shown in Figure 3, the trajectory of harmonic current can be observed in the case of OPF and OSF.

$$u_{WO} = \begin{cases} 0, & i_W > 0 \\ U_{dc}, & i_W < 0 \end{cases}, \text{ if } T_{61} = 1 \tag{4}$$

$$u_{WO} = 0, \text{ if } T_{61} = 0 \tag{5}$$

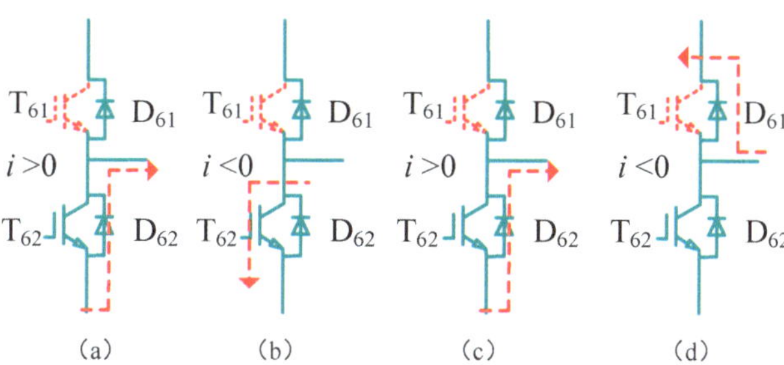



**Figure 2.** Current paths under different switch states and current directions (The red arrow shows the current flow direction): (**a**) $i_W > 0$, $T_{61} = 0$, and $T_{62} = 1$; (**b**) $i_W < 0$, $T_{61} = 0$, and $T_{62} = 1$; (**c**) $i_W > 0$, $T_{61} = 1$, and $T_{62} = 0$; and (**d**) $i_W < 0$, $T_{61} = 1$, and $T_{62} = 0$.

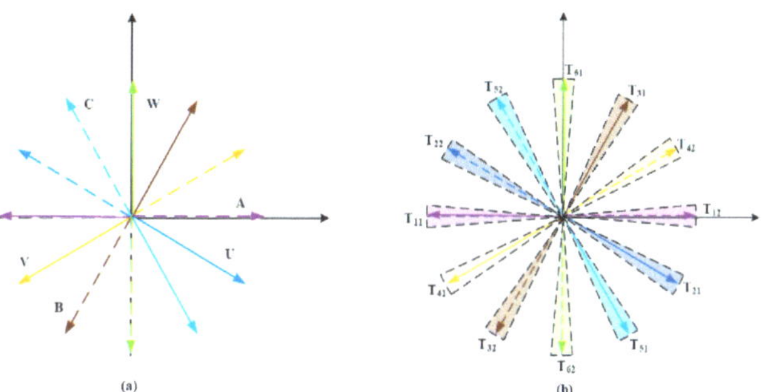



**Figure 3.** x–y plane current trajectories: (**a**) different open-circuit faults; (**b**) different switching tube of an inverter.

Similarly, this method can be applied to analyze various open-circuit fault scenarios. Figure 4. illustrates the fault phase currents for three types of faults: a single inverter open circuit, a single-phase open circuit, and a two-phase open circuit, when the drive system experiences a failure.

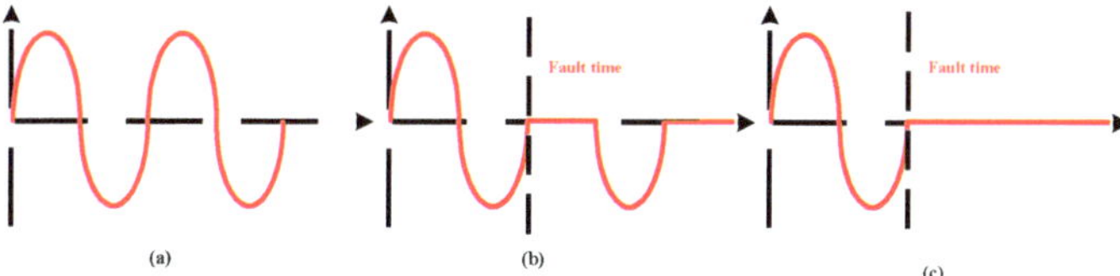



**Figure 4.** Current waveform under the same conditions. (**a**) Health; (**b**) single pipe open circuit; (**c**) single-phase open circuit.

## 4. Proposed Fault Diagnosis Method

In this section, we propose a novel method for diagnosing multiple open-circuit faults in the OPF system based on the dual d–q predicted current model, as depicted in Figure 5.

In this paper's electrical fault diagnosis, we present a two-step diagnostic scheme. Firstly, system faults are identified by monitoring the fault indicator $F_n$. Secondly, the phase sequence of open-circuit faults is determined through the fault phase indicator $D_N$. It is possible to locate the specific faulty transistor by evaluating the magnitude of the inverter fault indicator $T_M$ corresponding to inverter PINs 1 to 12.

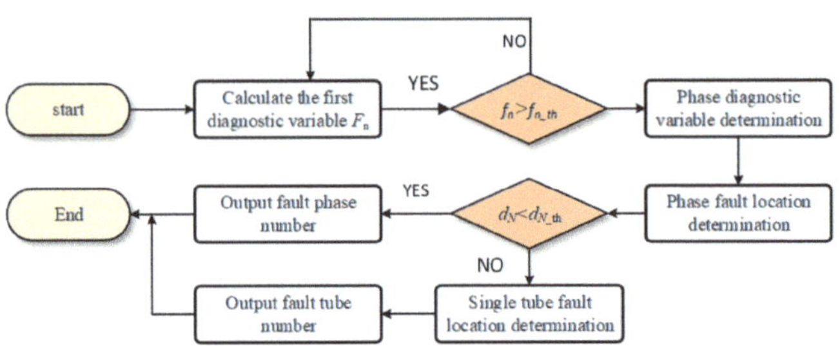


**Figure 5.** Flowchart of the diagnosis process.

### 4.1. Fault Current Model

Under normal operation, the predicted current in the (k + 1)th sampling interval can be derived from the current in the kth sampling interval according to Equation (3). If there is a fault, the predicted current in the (k + 1)th sampling interval predicted by the normal model will deviate from the feedback current. First, the correct predicted current in the fault operation needs to be calculated for fault diagnosis.

The stator voltage of a dual three-phase permanent magnet synchronous motor is expressed as

$$U_{6s} = R_{6s} I_{6s} + \frac{d\psi_{6s}}{dt} \tag{6}$$

$U_{6s}$, $I_{6s}$, $\psi_{6s}$, and $R_{6s}$ represent the six-phase stator voltage vector, phase current vector, magnetic flux linkage, and stator resistance, respectively.

Because of the neutral isolation, each winding can be treated as an independent winding. The phase-current relation can be changed as follows

$$\begin{cases} i_A + i_B + i_C = 0 \\ i_U + i_V + i_W = 0 \end{cases} \tag{7}$$

Assuming that the A-phase current will be forced to zero in case of an open-circuit fault in the A phase. According to Kirchhoff's law, the relationship between the B-phase current and the C-phase current can be expressed as follows

$$i_{B\_OA} = -i_{C\_OA} \tag{8}$$

According to Equations (6) and (8), the line voltage between phases B and C under A-phase OPF can be expressed as

$$U_{BC\_OA} = 2R_s i_{B\_OA} + 2L_d \frac{di_{B\_OA}}{dt} + \sqrt{3}\omega_e \cos\theta_e \psi_f \tag{9}$$

Through the discretization processing of Equation (3), it can be obtained that

$$U^k_{BC\_OA} = 2R_s i^k_{BC\_OA} + 2(i^{k+1}_{BC\_OA} - i^k_{BC\_OA})\frac{L_d}{T_s} + \sqrt{3}\omega_e \cos\theta_e \psi_f \tag{10}$$

According to Equation (10), the predicted B-phase current during the (k + 1)th sampling interval under A-phase OPF operation can be represented as

$$i^{k+1}_{\text{BC\_OA}} = \left( u^{k+1}_{\text{BC\_OA}} - \sqrt{3}\omega_e \cos\theta_e \psi_f - 2R_s i^{k}_{\text{BC\_OA}} \right)\frac{T_s}{2L_d} + i^{k}_{\text{BC\_OA}} \tag{11}$$

Thus, using Park transformation, the predictive current in the d1q1 axis of the OPF model under the A-phase OPF constraint can be derived from Equations (7) and (11), expressed as

$$\begin{cases} i^{k+1}_{\text{d1\_OA}} = \frac{\sqrt{3}}{3}\sin\theta\, i^{k+1}_{\text{B}} \\ i^{k+1}_{\text{q1\_OA}} = \frac{\sqrt{3}}{3}\cos\theta\, i^{k+1}_{\text{B}} \end{cases} \tag{12}$$

Similarly, when an open-circuit fault occurs in the U phase, the fault current constraint can be expressed as

$$i_{\text{V\_OU}} = -i_{\text{W\_OU}} \tag{13}$$

Under the constraint of the U-phase fault current, the predictive current in the d2–q2 axis of the OPF model under the U-phase OPF constraint can be derived

$$\begin{cases} i^{k+1}_{\text{d2\_OU}} = \frac{\sqrt{3}}{3}i^{k+1}_{\text{V}}\sin(\theta - \frac{\pi}{6}) \\ i^{k+1}_{\text{q2\_OU}} = \frac{\sqrt{3}}{3}i^{k+1}_{\text{V}}\cos(\theta - \frac{\pi}{6}) \end{cases} \tag{14}$$

Due to the spatial symmetry of the windings in a dual three-phase permanent magnet synchronous motor, the dual dq-axis predictive current model under the other phase's OPF constraints can be obtained by adjusting the angles according to Equations (12) and (14). In summary, the fault current model can be represented by the formulas (15) and (16)

$$\begin{cases} i^{k+1}_{\text{d1\_on}} = (-\omega^2 + \omega + 1)\frac{\sqrt{3}}{3}i^{k+1}_{n*}\sin(\theta + \omega\frac{\pi}{3}) \\ i^{k+1}_{\text{q1\_on}} = (-\omega^2 + \omega + 1)\frac{\sqrt{3}}{3}i^{k+1}_{n*}\cos(\theta + \omega\frac{\pi}{3}) \end{cases} \tag{15}$$

$$\begin{cases} i^{k+1}_{\text{d2\_om}} = \frac{\sqrt{3}}{3}i^{k+1}_{m*}\sin(\theta + \lambda\frac{\pi}{6}) \\ i^{k+1}_{\text{q2\_om}} = (-1)^{\lambda+1}\frac{\sqrt{3}}{3}i^{k+1}_{m*}\cos(\theta + \lambda\frac{\pi}{6}) \end{cases} \tag{16}$$

In the equation, the variables are as follows: $n$ = (A, B, C); $n^*$ = (B, A, A); $\omega$ = (0, 1, −1); $m$ = (U, V, W); $m^*$ = (V, U, U); $\lambda$ = (−1, 1, 0).

### 4.2. Determination of Fault Occurrence

According to the analysis of fault current in Section 3, the current trajectories of OPF and OSF have the same trajectory characteristics in the harmonic plane, so they can be used for fault determination. The harmonic predicted current vector mode is defined as

$$f_n = \sqrt{\left(i^{k+1}_x\right)^2 + \left(i^{k+1}_y\right)^2} \tag{17}$$

Define $f_n$ as an intermediate variable for fault detection. Under normal conditions, $i_x$ and $i_y$ are nearly zero, keeping $f_n$ close to zero. In the event of various open-circuit faults, the harmonic currents exhibit a predictable distribution, causing the magnitude of the vector to significantly deviate from zero. Consequently, monitoring changes in $f_n$ can enable rapid fault detection.

To reduce the effects of random noise in time series data, a moving average filter is used to smooth the signal waveform. Post-filtering, the filtered value $f_{n\_mean}$ is assigned to $f_n$ as a basis for fault detection in the algorithm.

$$f_{n\_mean} = \frac{1}{T}\int_{T_0}^{T_0+T} f_n(x)\,dx \tag{18}$$

In the experiment described in this paper, the index $f_n$ has a maximum error of about 0.2 A during normal operation. To avoid misdiagnosis and ensure sufficient safety margin, the threshold of the index $f_{n\_th}$ is set to 0.3 A. When $f_{n\_mean}$ is greater than the threshold, the system is determined to be in fault and the fault index $F_n = 1$ is output.

### 4.3. Identification of Faulty Phase

As shown in Figure 5, the first step of the fault diagnosis process, $F_n$, determines whether the system is at fault. The second step of the fault diagnosis process narrows the fault search range to specific faults, such as single inverter OSF and single-phase OPF.

Under fault, the feedback currents of the fault model of axis d1–q1 in the (k + 1)th sampling interval are almost the same as the predicted currents of axis d1–q1 of the normal model, which can be expressed as.

$$\begin{cases} i_{d1}^{k+1} = i_{d1\_ON}^{k+1} \\ i_{q1}^{k+1} = i_{q1\_ON}^{k+1} \end{cases} \tag{19}$$

According to Equation (20) the equation of residual currents of planar fault prediction current and health prediction current under the constraint of A- and U-phase OPF is obtained as follows

$$\begin{cases} \Delta i_{d1\_OA}^{k+1} = i_{d1\_OA}^{k+1} - i_{d1}^{k+1} \\ \Delta i_{q1\_OA}^{k+1} = i_{q1\_OA}^{k+1} - i_{q1}^{k+1} \end{cases} \tag{20}$$

Take the residual current vector mode and construct the intermediate variables $d_A$ and $d_U$ for phase diagnosis

$$d_A = \sqrt{\Delta i_{d1\_OA}^{k+1}{}^2 + \Delta i_{q1\_OA}^{k+1}{}^2} \tag{21}$$

The A-phase fault can be analyzed using Formula (21) to define two intermediate variables for phase location of winding faults, denoted as $d_N$ ($f = A, B, C; s = U, V, W$).

$$d_N = \begin{cases} d_f = \sqrt{\Delta i_{d1\_f}^2 + \Delta i_{q1\_f}^2} \\ d_s = \sqrt{\Delta i_{d2\_s}^2 + \Delta i_{q2\_s}^2} \end{cases} \tag{22}$$

The intermediate variables of fault phase location of two sets of windings are processed by sliding average filtering, respectively, and the filter $d_{N\_mean}$ is assigned to $d_N$ in the algorithm.

In the fault state, the residual model of the fault-predicted current is close to equal, and the residual vector modulus $d_A$ and $d_U$ under the fault is close to zero. In the healthy state of the system, $d_A$ will be greater than zero, and they are constant positive half-wave functions. Therefore, by observing the changes of $d_N$ and $d_{N\_th}$, the fault phase can be quickly positioned, and the waveforms of the intermediate variables $f_n$ and $d_N$ under different faults are shown in Table 1.

**Table 1.** Time-domain fault variable waveform under different faults.

| Intermediate Variable | Health | OSF | OPF |
|---|---|---|---|
| $f_n$ | | | |
| $d_N$ | | | |

### 4.4. Determination of Specific Fault Type

When the judgment condition satisfies the constraint conditions of $f_n > f_{n\_th}$ and $d_N < d_{N\_th}$, according to the judgment of the condition statement, the model will move to the positioning stage of the switch tube fault location. Here, "$f_n > f_{n\_th}$" refers to the first diagnostic variable indicator indicating that the system has a fault, while "$d_N < d_{N\_th}$" indicates that the phase fault indicator does not change and there is no open-circuit fault. At this point, it indicates that the system has a fault but not an open-circuit fault, that is the switch tube open-circuit fault. The model will execute the switch tube positioning program to determine the specific fault tube position.

The reference angle of the harmonic current vector at each phase fault can be described as follows

$$t_M* = \arctan\frac{i_x^{k+1}}{i_y^{k+1}} \tag{23}$$

The analysis in Section 3 reveals that the harmonic plane current trajectories undergo systematic variations when an OSF occurs, depending on the position of the faulty switch tube. Figure 6 presents a diagram illustrating the segmentation of current vector trajectories under a single-tube fault, (where $A^+$ denotes the upper switch of phase A and $A^-$ represents the lower switch of phase A).

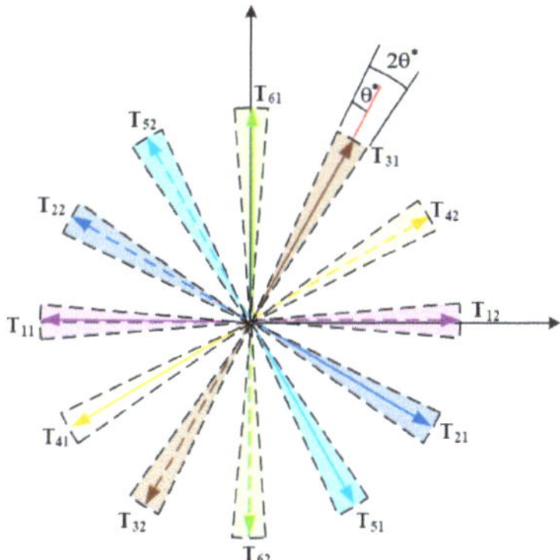


**Figure 6.** Position angle of ix–iy current track corresponding to inverter open-circuit fault.

To overcome the effects of non-ideal factors during the actual operation of the system, a safety threshold $\theta^*$ is added to the ideal position of the current trajectory. This ensures that once the diagnosis method calculates the current trajectory position as satisfying Equation (24), the position of the faulty switch can be determined. The selection of $\theta^*$ should be based on the actual operation of the system, and in this experiment, $\theta^*$ is chosen to be 5°.

$$t_M \in [(t_M^* - \theta*), (t_M^* + \theta*)] \tag{24}$$

As shown in Figure 6, the faulty tube position can be accurately identified by determining the position angle $t_M$ of the harmonic current trajectory within the corresponding error interval.

## 5. Experimental Verification

### 5.1. Experiment System

The diagnostic strategy's efficacy was validated through experiments conducted on a DTP-PMSM. The experimental platform is shown in Figure 7 The experimental platform demonstrates the functions and interrelationships of the various parts of the system.

Key motor parameters include: rated power of 4 kW; peak current of 3.5 A; rated speed of 1000 r/min; and stator resistance $R_s$ 1.45 Ω. The control system's core is constructed using TI TMS320F28379 DSP of USA and Intel's FPGA MAX-V chip of USA. During the experiment, a switching frequency of 10 kHz was employed, with load provided by the PMSM while vector control and fault diagnosis algorithms were implemented through DSP technology. Fault simulation was achieved by blocking corresponding pulse drive signals.

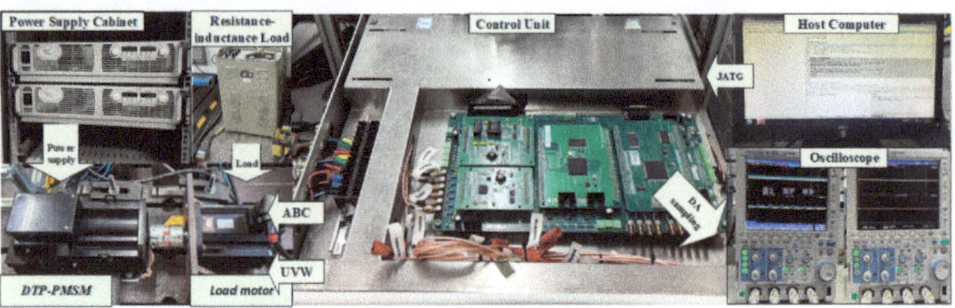


**Figure 7.** Experimental platform for DTP-PMSM control system.

### 5.2. Transient Performance Analysis

Experimental validations were conducted to assess the robustness of the proposed fault diagnosis scheme during transient operations, specifically evaluating speed and load response. Figure 8a depicts an acceleration experiment where a constant torque of 5 N·m was applied, increasing the speed from 400 r/min to 800 r/min within 0.5 s after achieving stable operation for a certain duration. Throughout the variable-speed process, transient step phenomena were observed in the intermediate variables. In the load variation experiment shown in Figure 8b, after accelerating and maintaining a stable speed of 800 r/min for a specific period, the load was abruptly increased to 8 N·m within 0.5 s due to loading effects. This led to an increase in phase current amplitude and intermediate variables, resulting in changes in motor speed and load; however, these variations did not affect fault diagnosis as protection strategies employed an undershoot threshold setting method. During the testing period, all intermediate variables consistently remained above their thresholds while the fault detection index $D_N$ consistently maintained a zero value, thereby validating both the effectiveness and stability of this diagnostic approach during transient operations.

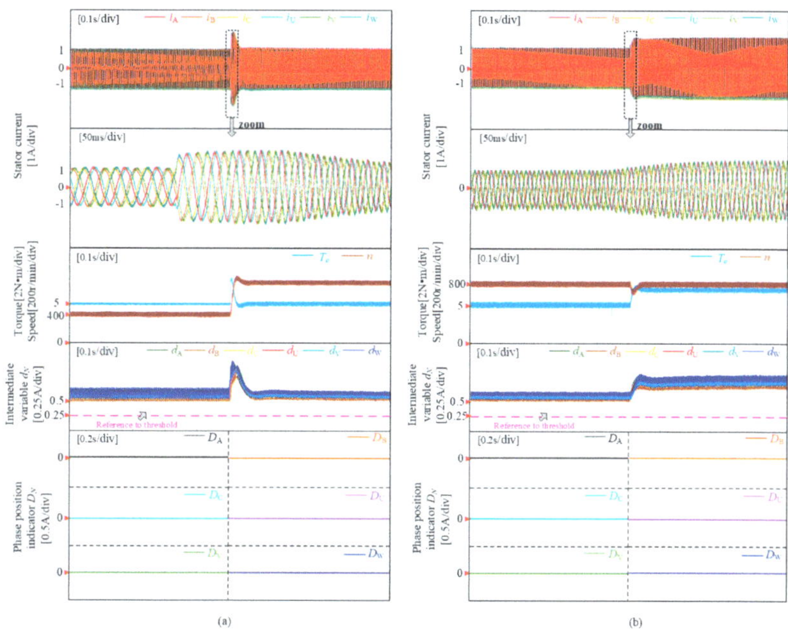

**Figure 8.** Health state transient waveform. (**a**) Acceleration under constant load; (**b**) constant speed loading.

## 5.3. Dynamic Performance Analysis

Figure 9 depicts the experiment of fault diagnosis in the PMSM drive system under open-circuit fault conditions in phase A. During normal operation, the phase A current exhibits a sinusoidal waveform with the fault occurrence index $f_n$ significantly below its threshold $f_{n\_th}$. However, when an open-circuit fault occurs in phase A, the current drops to zero and remains constant. At this point, the fault occurrence index briefly exceeds its threshold $f_{n\_th}$, resulting in the transition of the fault flag $F_n$ from 0 to 1 within 5 ms, indicating that a fault has occurred and completing the first diagnostic step. After an open-circuit fault occurs in phase A, intermediate variable $d_A$ rapidly decreases below its threshold and causes a surge of $D_A$ from 0 to 1 within 10 ms signifying the completion of the diagnostic process. Additionally, all five intermediate variables $d_A - d_V$ remain above their thresholds confirming no false positives during diagnosis.

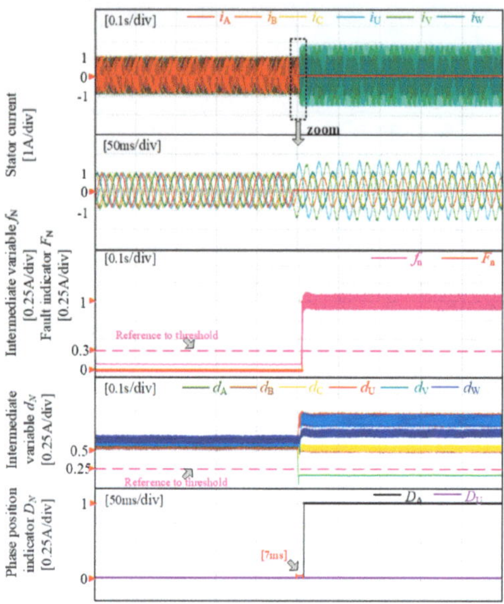

**Figure 9.** Fault diagnosis of phase—A OPF.

The fault diagnosis experiment of loading a U-phase open-circuit PMSM drive with an A-phase open-circuit fault is depicted in Figure 10. Under normal operating conditions, the fault occurrence index $f_n$ remains significantly lower than its threshold value $f_{n\_th}$. However, when the U-phase experiences OPF following the A-phase, throughout the entire process, the fault occurrence index $f_n$ consistently exceeds the threshold value $f_{n\_th}$, indicating that the initial step of diagnosis is always responsive. Before the U-phase fault, the phase indicator DA consistently remains at 1; after the fault occurs, $d_U$ rapidly decreases. Within 7 ms, the fault phase indicator $D_U$ transitions from 0 to 1, signifying completion of the diagnostic process. Throughout this process, four additional intermediate variables $d_B{\sim}d_C$ and $d_V{\sim}d_W$ all surpass their thresholds while indicators $d_B{\sim}d_C$ and $d_V{\sim}d_W$ remain zero; thus ensuring no misjudgment in this procedure.

The fault diagnosis experiment of the open-circuit fault in the PMSM drive system is illustrated in Figure 11, focusing on the A-phase switching tube $T_{12}$. Under normal operation, the A-phase current exhibits a sinusoidal waveform with an occurrence index $f_n$ significantly below the threshold $f_{n\_th}$. However, when $T_{12}$ experiences an open-circuit fault, the A-phase current drops to zero during the negative half cycle and approximates a sine waveform during the positive half cycle. Simultaneously, there is a momentary increase in the fault occurrence index $f_n$ above its threshold $f_{n\_th}$ value, leading to a transition of fault flag $F_n$ from 0 to 1 indicating completion of the initial diagnostic step. Furthermore, as evidenced by the phase positioning index $d_N$ surpassing its threshold value, it can be determined that this is a single inverter fault with the identified position. Upon confirming $T_{12}$ open-circuit fault condition, $T_M$ fault flag changes from 0 to 7 within 28 ms signifying the completion of the diagnostic process and determination of the seventh pin (i.e., lower bridge arm of A-phase) as a faulty location.

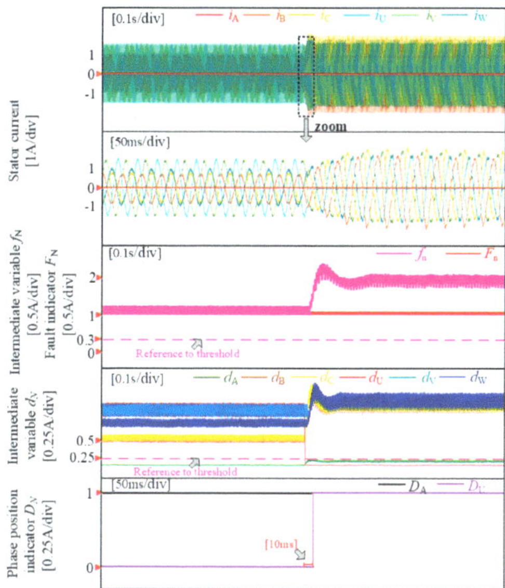

**Figure 10.** Fault diagnosis of phase—A and U OPF.

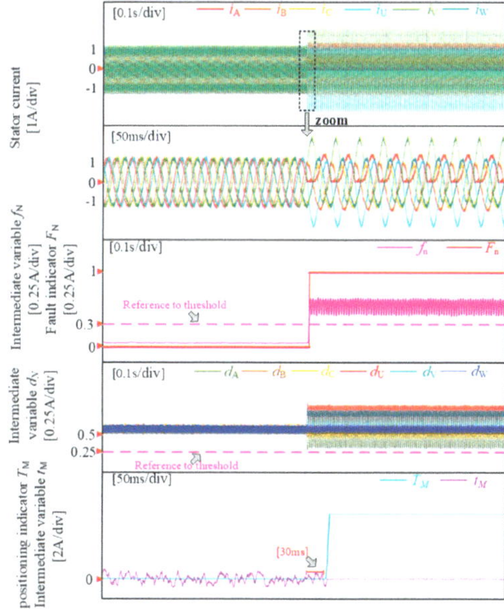

**Figure 11.** Fault diagnosis of SA1–OSF.

## 5.4. Parametric Disturbance Analysis

The stability of the diagnostic strategy is verified by introducing a 20% error separately in the d–q axis inductance and stator resistance and observing its impact on the diagnostic results. As depicted in Figure 12, parameter deviations cause minimal fluctuation in the waveform of intermediate variables during fault occurrence, which is significantly smaller than the maximum error of 0.1 A under normal conditions. The diagnostic intermediate variables $f_n$ are much smaller than the threshold $f_{n\_th}$, while the phase diagnostic intermediate variables $d_n$ are much larger than the threshold $d_{N\_th}$. Based on this analysis, it can be concluded that when parameter errors are introduced during healthy motor operation, the diagnostic algorithm performs well, and intermediate variable values remain considerably lower than their preset thresholds $f_{n\_th}$ and $d_{N\_th}$. This demonstrates that parameter disturbances do not affect diagnosis.

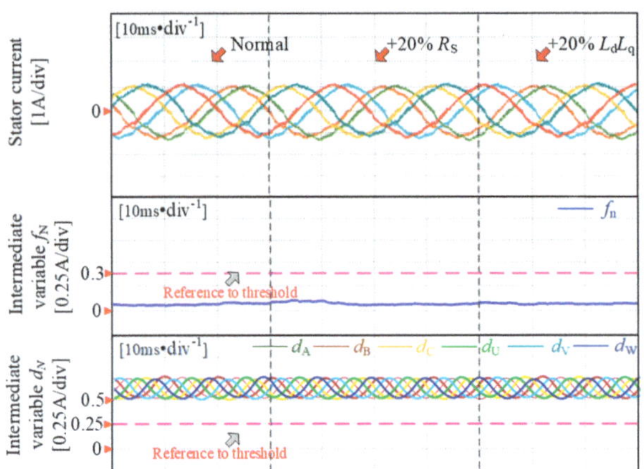


**Figure 12.** Comparison of parameter disturbance.

## 6. Conclusions

This study addresses the problem of incorrect fault diagnoses that frequently occur in a dual three-phase permanent magnet synchronous motor during load steps, idling, or at low load. An open-circuit fault diagnosis strategy is proposed to improve robustness. The theoretical analysis and experimental validation confirm the results:

(1) This strategy can diagnose three types of open-circuit faults in the DTP-PMSM drive system, including 12 fault types for OSF and 21 fault types for single-phase and two-phase OPF, making a total of 33 fault types. And, it can also diagnose single-phase open faults and two-phase open faults simultaneously.

(2) The strategy effectively overcomes the issue of misjudgment in load mutation encountered by traditional methods, The empirical verification has confirmed the demonstrating high reliability and robustness in diagnosing open-circuit faults. The diagnosis time for OPF and OSF in the system is within 10 ms and 30 ms, respectively.

(3) After simplifying the calculation process, the strategy of this paper achieves that there is no interference between the different diagnostic variables in the compatible diagnosis of multiphase open-circuit faults. Furthermore, this strategy does not require additional hardware support and can perform fault diagnosis quickly and accurately even when operating points, control strategies or drive parameters change.

(4) This strategy employs the approach of utilizing predicted current extraction to construct fault models, which acts as a link between fault diagnosis and fault-tolerant control based on predicted current, providing certain support for future research on a dual three-phase permanent magnet synchronous motor's fault-tolerant control and laying part of the foundation.

**Author Contributions:** Conceptualization, Q.G.; data curation, W.D.; formal analysis, G.Z.; funding acquisition, Z.Z. and Q.G.; investigation, X.J.; methodology, Z.Z.; project administration, Q.G.; resources, Q.G.; software, X.J.; supervision, G.Z. and Q.G.; validation, X.J.; visualization, W.D.; writing—original draft, W.D. All authors have read and agreed to the published version of the manuscript.

**Funding:** This research was supported in part by the National Natural Science Foundation of China under Grant 52077154, in part by the Natural Science Foundation of Tianjin Municipality under Grant 23JCYBJC00290, and in part by the Zhejiang Provincial Natural Science Foundation of China under Grant LY24E070003.

**Institutional Review Board Statement:** Not applicable.

**Informed Consent Statement:** Not applicable.

**Data Availability Statement:** The original contributions presented in the study are included in the article, further inquiries can be directed to the corresponding author.

**Conflicts of Interest:** The authors declare no conflict of interest.

## References


1. Levi, E. Advances in converter control and innovative exploitation of additional degrees of freedom for multiphase machines. *IEEE Trans. Ind. Electron.* **2015**, *63*, 433–448.
2. Wu, G.; Huang, S.; Wu, Q.; Rong, F.; Zhang, C.; Liao, W. Robust Predictive Torque Control of N*3-Phase PMSM for High-PowerTraction Application. *IEEE Trans. Power Electron.* **2020**, *35*, 10799–10809.
3. Riaz, S.; Qi, R.; Tutsoy, O. A novel adaptive PD-type iterative learning control of the PMSM servo system with the friction uncertainty in low speeds. *PLoS ONE* **2023**, *18*, 1932–6203. [CrossRef]
4. Ping, Z.; Wang, T.; Huang, Y.; Wang, H.; Lu, J.G.; Li, Y. Internal model control of PMSM position servo system: Theory and experimental results. *IEEE Trans. Ind. Inform.* **2020**, *16*, 2202–2211.
5. Li, L.N.; Zhu, G.J. Electromagnetic-thermal-stress efforts of stator-casing grease buffers for permanent magnet driving motors. *IEEE Trans. Ind. Appl.* **2024**, *60*, 1268–1276. [CrossRef]
6. Kommuri, S.K.; Defoort, M.; Karimi, H.R.; Veluvolu, K.C. A robust observer-based sensor fault-tolerant control for PMSM in electric vehicles. *IEEE Trans. Ind. Electron.* **2016**, *63*, 7671–7681.
7. Choi, U.M.; Blaabjerg, F.; Lee, K.B. Study and handling methods of power IGBT module failures in power electronic converter systems. *IEEE Trans. Power Electr.* **2014**, *30*, 2517–2533.
8. Wang, X.; Wang, Z.; Xu, Z. Comprehensive diagnosis and tolerance strategies for electrical faults and sensor faults in dual three-phase PMSM drives. *IEEE Trans. Power Electron.* **2018**, *34*, 6669–6684.
9. Choi, U.M.; Lee, K.B.; Blaabjerg, F. Diagnosis and tolerant strategy of an open-switch fault for T-type three-level inverter systems. *IEEE Trans. Ind. Appl.* **2013**, *50*, 495–508.
10. Choi, U.M.; Jeong, H.G.; Lee, K.B. Method for detecting an open-switch fault in a grid-connected NPC inverter system. *IEEE Trans. Power Electron.* **2011**, *27*, 2726–2739.
11. Li, H.; Guo, Y.; Xia, J.; Zhang, X. Open-circuit fault diagnosis for a fault-tolerant three-level neutral-point-clamped STATCOM. *IET Power Electron.* **2019**, *12*, 810–816.
12. Lee, J.S.; Lee, K.B.; Blaabjerg, F. Open-switch fault detection method of a back-to-back converter using NPC topology for wind turbine systems. *IEEE Trans. Ind. Appl.* **2014**, *51*, 325–335.
13. Shi, T.; He, Y.; Deng, F.; Shi, L. Online diagnostic method of open-switch faults in PWM voltage source rectifier based on instantaneous AC current distortion. *IET Electric Power Appl.* **2018**, *12*, 447–454.
14. Shi, T.; He, Y.; Wang, T.; Deng, F. An improved open-switch fault diagnosis technique of a PWM voltage source rectifier based on current distortion. *IEEE Trans. Power Electron.* **2019**, *34*, 12212–12225.
15. Wu, F.; Zhao, J. Current similarity analysis-based open-circuit fault diagnosis for two-level three-phase PWM rectifier. *IEEE Trans. Power Electron.* **2016**, *32*, 3935–3945.
16. Guo, H.; Guo, S.; Xu, J.; Tian, X. Power switch open-circuit fault diagnosis of six-phase fault tolerant permanent magnet synchronous motor system under normal and fault-tolerant operation conditions using the average current Park's vector approach. *IEEE Trans. Power Electron.* **2020**, *36*, 2641–2660.
17. Xu, J.; Tian, X.; Jin, W. Pwm harmonic-current-based interturn short-circuit fault diagnosis for the aerospace FTPMSM system even in the fault-tolerant operation condition. *IEEE Trans. Power Electron.* **2023**, *38*, 5432–5441.
18. Wang, B.; Li, Z.; Bai, Z.; Krein, P.T.; Ma, H. A voltage vector residual estimation method based on current path tracking for T-type inverter open-circuit fault diagnosis. *IEEE Trans. Power Electron.* **2021**, *36*, 13460–13477.
19. De Araujo Ribeiro, R.L.; Jacobina, C.B.; Da Silva, E.R.C.; Lima, A.M.N. Fault detection of open-switch damage in voltage-fed PWM motor drive systems. *IEEE Trans. Power Electron.* **2003**, *18*, 587–593.
20. Wu, X.; Chen, C.Y.; Chen, T.F.; Cheng, S.; Mao, Z.H.; Yu, T.J.; Li, K. A fast and robust diagnostic method for multiple open-circuit faults of voltage-source inverters through line voltage magnitudes analysis. *IEEE Trans. Power Electron.* **2019**, *35*, 5205–5220.
21. Hang, J.; Zhang, J.; Cheng, M.; Ding, S. Detection and discrimination of open-phase fault in permanent magnet synchronous motor drive system. *IEEE Trans. Power Electron.* **2015**, *31*, 4697–4709.
22. Yin, H.; Chen, Y.; Chen, Z.; Li, M. Adaptive fast fault location for open-switch faults of voltage source inverter. *IEEE Trans. Circ Syst. 1 Reg Papers.* **2021**, *68*, 3965–3974.
23. Zhang, W.; He, Y. A hypothesis method for t-type three-level inverters open-circuit fault diagnosis based on output phase voltage model. *IEEE Trans Power Electron.* **2022**, *37*, 9718–9732. [CrossRef]
24. Cheng, Y.; Sun, Y.; Li, X. Active common-mode voltage-based open-switch fault diagnosis of inverters in im-drive systems. *IEEE Trans Ind Electron.* **2021**, *68*, 103–115. [CrossRef]

*World Electric Vehicle Journal*

*Article*

# Direct Torque Control of Dual Three-Phase Permanent Magnet Synchronous Motors Based on Master–Slave Virtual Vectors

**Qiang Geng [1], Ziteng Qin [1], Xuefeng Jin [1], Guozheng Zhang [1] and Zhanqing Zhou [2],***

[1]  Tianjin Key Laboratory of Intelligent Control of Electrical Equipment, Tiangong University, Tianjin 300387, China; gengqiang@tju.edu.cn (Q.G.); 2131070894@tiangong.edu.cn (Z.Q.); jinxuefeng@tiangong.edu.cn (X.J.); zhanggz@tju.edu.cn (G.Z.)

[2]  Advanced Electrical Equipment Innovation Center, Zhejiang University, Hangzhou 311107, China

*  Correspondence: zhouzhanqing@tiangong.edu.cn

**Abstract:** In order to further reduce the torque, flux-linkage fluctuation, and current harmonic content of dual three-phase permanent magnet synchronous motors, this paper proposes a direct torque control strategy combined with a master–slave virtual vector duty cycle assignment. Two types of virtual voltage vectors with different amplitudes are used to form a harmonic suppression switching table. The virtual vectors are classified into master and slave virtual vectors according to the degree of influence on the torque and the flux-linkage. Then, the duty cycle of the master and slave virtual vectors is recalculated and allocated through the evaluation function to achieve accurate control of the torque and the flux-linkage. Finally, the switching sequences of the master and slave virtual vectors that act together in one control cycle are rearranged into a symmetrical waveform. It is experimentally verified that the phase current THD of the proposed strategy is reduced by 69.4%, the 5th and 7th current harmonics content is significantly reduced, and the torque fluctuation and flux-linkage fluctuation can also be effectively suppressed, which provides better dynamic performance and steady-state performance.

**Keywords:** dual three-phase permanent magnet synchronous motor; direct torque control; master–slave virtual vectors; duty cycle distribution; switch sequence correction

**Citation:** Geng, Q.; Qin, Z.; Jin, X.; Zhang, G.; Zhou, Z. Direct Torque Control of Dual Three-Phase Permanent Magnet Synchronous Motors Based on Master–Slave Virtual Vectors. *World Electr. Veh. J.* **2024**, *15*, 199. https://doi.org/10.3390/wevj15050199

Academic Editor: Paulo J. G. Pereirinha

Received: 18 March 2024
Revised: 26 April 2024
Accepted: 1 May 2024
Published: 4 May 2024

## 1. Introduction

With the increasing demand for the power and safety of motor systems in engineering applications such as aerospace and electric vehicles, the research on multiphase motors with high power and high reliability has attained significant proportions in recent years. Among various multiphase drives, the dual three-phase permanent magnet synchronous motor (DTP-PMSM) drives have the advantages of both multiphase motor drives and permanent magnet motor drives. Compared to three-phase drives, the DTP-PMSM drives offer superior features of reduced torque ripple and remarkable fault tolerance capability [1–3].

The direct torque control (DTC) scheme has a simpler structure and faster dynamic response in comparison with field-oriented control (FOC). Meanwhile, for DTP-PMSMs, the six-phase inverter that drives the motor operation can generate more voltage vectors, which provides a rich vector control set for the DTC scheme and enhances the flexibility of the control strategy, but also increases the difficulty of the algorithm design [4,5]. In addition, it is noted that only closed-loop control of the fundamental subspace will cause a larger harmonic current due to the relatively small impedance of harmonic subspace in the mathematical model of vector space decomposition (VSD) [6,7].

The voltage vectors controlling the DTP-PMSM drives contain 60 effective vectors and 4 zero vectors, which can be divided into four groups according to the magnitude of the vectors. So as to decrease the complexity of the switching table and increase the voltage utilization, the traditional DTC switching table of DTP-PMSMs only selects the

12 large vectors with the largest magnitude on the fundamental subspace. For the sake of solving the problem of current harmonics in a large vector-based DTC scheme, the modified intermediate vectors are applied to suppress the 5th and 7th harmonics in the harmonic subspace. The modified vector was centralized for simplifying the hardware implementation [8]. In some of the literature [9–12], the non-harmonic voltage vector switching table is applied to reduce the current harmonics. Moreover, the high-order torque hysteresis controller is used to divide the hysteresis width into several adjustment intervals, which can suppress the torque fluctuation and steady-state error, but the voltage utilization of this strategy will be reduced to a certain extent. To suppress the torque fluctuation of the DTC of DTP-PMSMs, some scholars have proposed vector space modulation technology. Scholars propose a direct torque control strategy of DTP-PMSMs based on vector space modulation, which not only reduces harmonic current and torque fluctuation but also ensures constant switching frequency [13]. In [14], a hybrid DTC strategy suitable for DTP-PMSMs is proposed. The strategy adopts different torque control modes in dynamic and steady-state modes, respectively, and an observer is designed to realize the tracking control of the flux-linkage. Some scholars also combine DTC with model predictive control (MPC) to reduce the predicted voltage vectors [15,16]. In [17], based on 36 effective vectors, the two-step table lookup method is proposed to reduce the calculated vectors. Firstly, 13 vectors are screened out in the fundamental subspace by the positive and negative of the torque error, and then the vectors are further reduced according to the principle of reducing the flux of the harmonic subspace, which simplifies the computational process. In [18], an enhanced DTC strategy based on discrete virtual vectors is proposed to improve the steady-state performance of PMSMs. The proposed strategy maintains the simple control structure and ideal dynamic performance of the DTC strategy and improves the steady-state performance by reducing torque ripple and current distortion. The experiment verifies the effectiveness of the proposed DTC strategy. In [19], the article considers the potential voltage imbalance in the DC link between two DC voltage source inverters of DTP-PMSMs. In order to solve these problems, a DTC strategy considering unbalanced DC side voltage is proposed in this study to improve the performance of DTP-PMSMs using virtual vectors.

Compared with vector control strategies, traditional DTC control saves the complicated coordinate transformation, and can realize the fast response adjustment of the motor by controlling the torque. This method is simple in structure and is not sensitive to the parameters of the motor and external influences. However, in order to ensure the fast dynamic performance of the motor torque, the large vectors are selected for voltage synthesis. The large vector of the $\alpha$-$\beta$ subspace will generate a large harmonic current in the harmonic subspace, so the traditional DTC control strategy will have some problems such as a large torque ripple.

Aiming to address the limitations of traditional DTC control for DTP-PMSM drives, a master–slave virtual vector direct torque control strategy based on duty cycle allocation is put forward in this paper to enhance the accuracy of torque/flux-linkage control accuracy and suppress current harmonics. Firstly, different methods are employed to synthesize the virtual vector to suppress current harmonics under various motor operating conditions. Then, considering the impact of torque and flux-linkage, the synthesized virtual vector is divided into master and slave virtual vectors. The duty cycle of each virtual vector is calculated and redistributed based on specific deviations in torque and flux-linkage, enabling precise control over both parameters. Finally, the switch sequence generated by the master and slave virtual vectors is modified to achieve a symmetrical waveform output for motor control. The simulation and experimental results show that compared with the traditional DTC control strategy, the proposed strategy can not only effectively suppress the flux-linkage and torque ripple under different operating conditions but also reduce the difficulty of hardware implementation and ensure the effectiveness of harmonic suppression through the reallocation of the master–slave virtual vector duty cycle.

## 2. Mathematical Model of Dual Three-Phase PMSMs

The stator of a DTP-PMSM consists of two sets of Y-connected three-phase symmetrical winding, which are spatially separated by an electrical angle of 30° and are driven by a six-phase voltage source inverter, as shown in Figure 1.

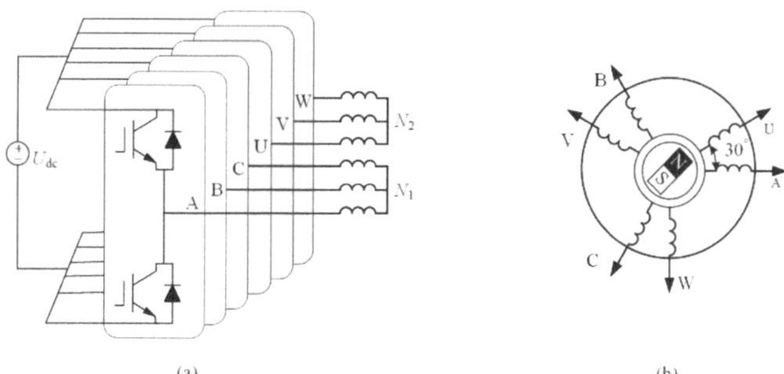

**Figure 1.** Dual three-phase PMSM drive system diagram: (**a**) six-phase inverter; (**b**) dual three-phase PMSM.

The static coordinate transformation matrix of DTP-PMSMs is shown in (1).

$$T_{6s/2s} = \frac{1}{3}\begin{bmatrix} 1 & -\frac{1}{2} & -\frac{1}{2} & \frac{\sqrt{3}}{2} & -\frac{\sqrt{3}}{2} & 0 \\ 0 & \frac{\sqrt{3}}{2} & -\frac{\sqrt{3}}{2} & \frac{1}{2} & \frac{1}{2} & -1 \\ 1 & -\frac{1}{2} & -\frac{1}{2} & -\frac{\sqrt{3}}{2} & \frac{\sqrt{3}}{2} & 0 \\ 0 & -\frac{\sqrt{3}}{2} & \frac{\sqrt{3}}{2} & \frac{1}{2} & \frac{1}{2} & -1 \\ 1 & 1 & 1 & 0 & 0 & 0 \\ 0 & 0 & 0 & 1 & 1 & 1 \end{bmatrix} \tag{1}$$

According to the VSD theory, the mathematical model in the static coordinate system of DTP-PMSMs can be decoupled into three two-dimensional orthogonal subspaces, respectively, $\alpha$-$\beta$, $x$-$y$, and $o_1$-$o_2$. The components on the $\alpha$-$\beta$ subspace comprise the fundamental and harmonic components with orders of 12 m $\pm$ 1 (m = 1, 3, 5, $\cdots$), which are related to torque generation. The components on the $x$-$y$ subspace comprise the harmonic components with orders of 6 m $\pm$ 1 (m = 1, 3, 5, $\cdots$), which do not participate in torque generation. The components on the $o_1$-$o_2$ subspace are the zero-sequence components, which comprise the harmonic components with orders of 6 m $\pm$ 3 (m = 1, 3, 5, $\cdots$).

The transformation matrix of the rotating coordinate system of DTP-PMSMs is shown in (2).

$$T_{s/r} = \begin{bmatrix} \cos\theta & \sin\theta & 0 \\ -\sin\theta & \cos\theta & 0 \\ 0 & 0 & I_A \end{bmatrix} \tag{2}$$

When the neutral points of the two winding sets are isolated from each other, the zero-sequence subspace can be ignored, then the voltage and torque equations in the rotating coordinate system are as follows, respectively:

$$\begin{bmatrix} u_d \\ u_q \end{bmatrix} = \begin{bmatrix} R_s & \omega L_q \\ \omega L_d & R_s \end{bmatrix}\begin{bmatrix} i_d \\ i_q \end{bmatrix} + \begin{bmatrix} L_d \\ L_q \end{bmatrix}\begin{bmatrix} n_p i_d \\ n_p i_q \end{bmatrix} + \begin{bmatrix} 0 \\ \omega\psi_s \end{bmatrix} \tag{3}$$

$$\begin{bmatrix} u_x \\ u_y \end{bmatrix} = \begin{bmatrix} R_s & 0 \\ 0 & R_s \end{bmatrix}\begin{bmatrix} i_x \\ i_y \end{bmatrix} + L_z\begin{bmatrix} n_p i_x \\ n_p i_y \end{bmatrix} \tag{4}$$

$$\begin{bmatrix} \psi_d \\ \psi_q \\ \psi_x \\ \psi_y \end{bmatrix} = \begin{bmatrix} L_d & 0 & 0 & 0 \\ 0 & L_q & 0 & 0 \\ 0 & 0 & L_z & 0 \\ 0 & 0 & 0 & L_z \end{bmatrix} \begin{bmatrix} i_d \\ i_q \\ i_x \\ i_y \end{bmatrix} + \begin{bmatrix} \psi_f \\ 0 \\ 0 \\ 0 \end{bmatrix} \tag{5}$$

$$T_e = 3n_p(\psi_d i_q - \psi_q i_d) \tag{6}$$

where $u_d$, $u_q$ and $i_d$, $i_q$ represent the $dq$-axes stator voltages and currents, respectively; $u_x$, $u_y$ and $i_x$, $i_y$ represent the $xy$-axes stator voltages and currents, respectively; $L_d$, $L_q$, and $L_z$ denote the $dq$-axes inductances and the leakage inductance, respectively; $\psi_s$ is the stator flux-linkage; $\theta$ is the rotor position; $T_e$ is the electromagnetic torque; $n_p$ is the number of pole pairs; and $\omega$ is the electric angular velocity.

### 3. Traditional Direct Torque Control Scheme

The traditional DTC block diagram of DTP-PMSMs is shown in Figure 2 [20,21]. The electromagnetic torque $T_e$, flux-linkage amplitude $|\psi_s|$, and flux-linkage angle $\theta_s$ are obtained through the torque and flux observer on the $\alpha$-$\beta$ subspace. The error between the observed value and the given value is transmitted to the hysteresis comparator to obtain the increase and decrease signal of torque flux-linkage.

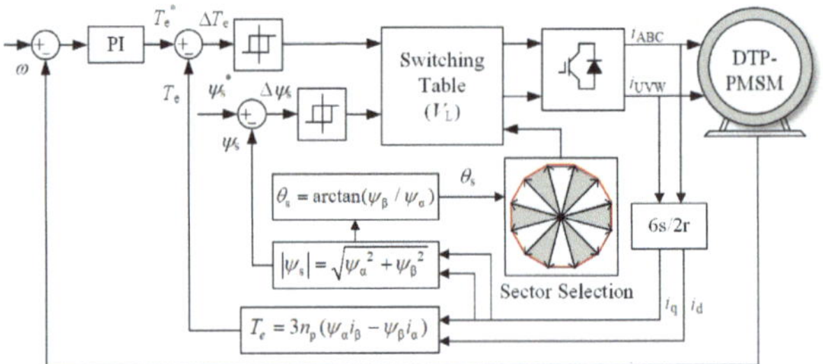

**Figure 2.** Block diagram of direct torque control for dual three-phase PMSMs.

A DTP-PMSM is driven by a six-phase inverter, where each switching state combination corresponds to a voltage vector in the $\alpha$-$\beta$ and $x$-$y$ subspaces, with $2^6 = 64$ different spatial voltage vectors, including 4 zero vectors and 60 effective vectors. The voltage vector distribution of the $\alpha$-$\beta$ and $x$-$y$ subspaces is shown in Figure 3. Numbers of the basic voltage vectors in the figure are, respectively, converted to octal numbers according to the order of bridge arms of ABC and UVW. Effective vectors can be divided into four groups: large vectors, medium–large vectors, medium vectors, and small vectors according to their different amplitudes on the $\alpha$-$\beta$ subspace. The amplitudes of voltage vectors of each group are $|V_L| = 0.644U_{dc}$, $|V_{ML}| = 0.471U_{dc}$, $|V_M| = 0.333U_{dc}$, and $|V_S| = 0.173U_{dc}$.

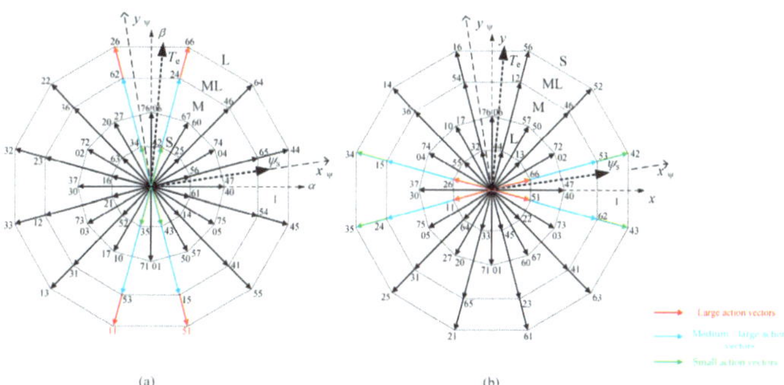

**Figure 3.** Dual three-phase PMSM voltage vector distribution diagram: (**a**) $\alpha$-$\beta$ subspace; and (**b**) $x$-$y$ subspace.

In order to reduce the complexity of the DTC switching table, the traditional strategy selects the twelve outermost vectors $V_L$ of $\alpha$-$\beta$ subspace. Taking the stator flux-linkage in sector I as an example, from the projection of each vector on the flux-linkage axis in Figure 3a, the vectors for torque increase are $V_{64}$, $V_{66}$, $V_{26}$, and $V_{22}$, where the vectors that have the greatest influence on torque increase are $V_{66}$ and $V_{26}$. Similarly, the vectors for torque reduction are $V_{55}$, $V_{51}$, $V_{11}$, and $V_{13}$, where the ones that have the greatest influence on torque reduction are $V_{66}$ and $V_{26}$. Therefore, according to the fact that the two conditions of the increase and decrease of the torque and the flux-linkage are satisfied at the same time, there will always be an effective vector that is selected in the switching table. Taking sector I as an example, the direct torque switching table for DTP-PMSMs is shown in Table 1.

**Table 1.** Sector I direct torque control switch table.

| Torque Error | Flux-Linkage Error | Voltage Vector |
| :---: | :---: | :---: |
| >0 | >0 | $V_{66}$ |
| <0 | >0 | $V_{51}$ |
| >0 | <0 | $V_{26}$ |
| <0 | <0 | $V_{11}$ |

## 4. Master–Slave Virtual Vector Direct Torque Control Strategy Based on Duty Cycle Allocation

### 4.1. Virtual Vector Synthesis Method

The traditional DTC strategy only considers the voltage vector effect in the $\alpha$-$\beta$ subspace and does not consider the harmonics caused by the vector in the $x$-$y$ subspace, so there is a large harmonic current. To solve the above problems, two kinds of virtual vector synthesis methods are proposed, which can control the voltage vector amplitude to zero in the harmonic subspace, so as to suppress current harmonics. The two kinds of virtual vectors can make full use of the rich voltage vectors set of DTP-PMSM and can also increase the control accuracy of motor torque under different states because of the different amplitude of the two kinds of virtual vectors.

As shown in Figure 3, $V_L$, $V_{ML}$, and $V_S$ have the same direction in the $\alpha$-$\beta$ subspace but opposite directions in the $x$-$y$ subspace. Therefore, the amplitude of the voltage vectors on the harmonic subspace can be controlled to zero by adjusting the action time. Let the action time of $V_L$, $V_{ML}$, and $V_S$ be $t_1$, $t_2$, and $t_3$, respectively. The first kind of virtual vector is synthesized via large vectors $V_L$ and medium–large vectors $V_{ML}$ in the same direction and satisfies the following relation:

$$\begin{cases} |VV_{n\alpha\beta}| = \frac{t_1|V_L|+t_2|V_{ML}|}{T_s} \ (n=1,\ldots,12) \\ |VV_{nxy}| = \frac{t_1|V_L|-t_2|V_{ML}|}{T_s} = 0 \ (n=1,\ldots,12) \end{cases} \tag{7}$$

The operation time $t_1$ and $t_2$ of large and medium–large vectors can be obtained from Equation (6) as follows:

$$\begin{cases} t_1 = 0.73T_s \\ t_2 = 0.27T_s \end{cases} \tag{8}$$

The second kind of virtual vector is synthesized via medium–large vectors $V_{ML}$ and small vectors $V_S$ in the same direction:

$$\begin{cases} |VV_{n\alpha\beta}| = \frac{t_2|V_{ML}|+t_3|V_S|}{T_s} \ (n = 13,\dots,24) \\ |VV_{nxy}| = \frac{t_2|V_{ML}|-t_3|V_S|}{T_s} = 0 \ (n = 13,\dots,24) \end{cases} \tag{9}$$

Similarly, the action time $t_2$ and $t_3$ of medium–large and small vectors can be solved as follows:

$$\begin{cases} t_2 = 0.58T_s \\ t_3 = 0.42T_s \end{cases} \tag{10}$$

The first kind of virtual vector voltage amplitude is $0.597U_{dc}$, which is very close to the $V_L$ amplitude. Under its action, the motor has larger torque and flux-linkage change, and the dynamic performance is better. The second kind of virtual vector voltage amplitude is $0.345U_{dc}$, about half of the $V_L$ amplitude, which can reduce torque fluctuations and has better steady-state performance, but it is not suitable for larger transient adjustment processes.

In the traditional strategy, regardless of the torque error value, the switching table always outputs a fixed amplitude vector, but this paper chooses the corresponding synthesis method according to the different operating states of the motor. The first kind of virtual vector $VV_{1-12}$ is used in the transient adjustment with a large error, and the second kind of virtual vector $VV_{13-24}$ is used in the steady adjustment with a small error. The distribution of two kinds of virtual vectors in the $\alpha$-$\beta$ subspace is shown in Figure 4.

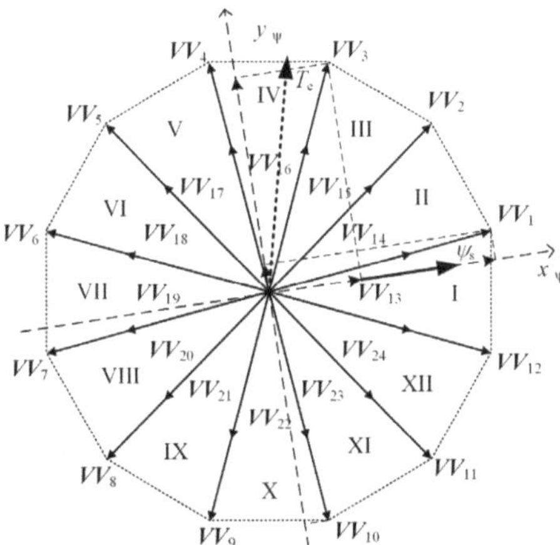

**Figure 4.** $\alpha$-$\beta$ subspace virtual vector distribution.

### 4.2. Master–Slave Virtual Vector Determination and Duty Cycle Allocation

The two types of virtual vectors of DTP-PMSMs are projected on the stator flux-linkage rotating coordinate system $x_\psi$-$y_\psi$, and the torque evaluation function $\lambda_T$, flux evaluation function $\lambda_\psi$, and back electromotive force evaluation function $\lambda_e$ are defined as follows [22]:

$$\begin{cases} \lambda_T = \frac{V_{y\psi}}{2/3U_{dc}} \\ \lambda_\psi = \frac{V_{x\psi}}{2/3U_{dc}} \\ \lambda_e = \frac{\omega|\psi_s|}{2/3U_{dc}} \end{cases} \tag{11}$$

where $V_{x\psi}$ and $V_{y\psi}$ are the components of the virtual voltage vector in the stator flux-linkage coordinate system $x_\psi$-axis and $y_\psi$-axis, respectively. The flux-linkage position angle $\theta_s$ determines the magnitude of the torque and flux-linkage evaluation function, while the back electromotive force evaluation function is related to the electric angular velocity $\omega$.

In order to enhance the torque control effect of DTP-PMSMs, the selection of voltage vectors is no longer limited to only 12 fixed directions. Depending on the sector in which the stator flux is located, the master virtual vector is chosen based on different evaluation functions for each voltage vector, considering its greater influence on torque. The slave virtual vector is selected based on its greater influence on flux-linkage. Both virtual vectors comply with the switching table requirements for torque and flux-linkage increases or decreases; however, their effects differ. Additionally, under different operating states, distinct types of virtual vectors should be selected.

For example, when the stator flux-linkage is in sector I, $\Delta T_e$ and $\Delta \psi_s$ are greater than zero. In the transient adjustment of the motor, the evaluation functions corresponding to all virtual vectors are calculated. The first kind of virtual vector $VV_{1-3}$ can increase the flux-linkage as well as the torque. The virtual vector $VV_3$ with the largest increase in torque is selected as the master virtual vector, and the virtual vector $VV_1$ with the largest increase in flux-linkage is selected as the slave virtual vector. Similarly, the second kind of virtual vector $VV_{15}$ is selected as the master virtual vector and vector $VV_{13}$ as the slave virtual vector during the steady state adjustment of the motor. According to the above theory, the master–slave virtual vectors satisfying the corresponding conditions can be selected in each sector.

The selected master–slave virtual vectors need to act together in a control period according to a certain duty cycle, and the torque and flux-linkage evaluation function can more accurately represent the degree of influence of the selected master–slave virtual vectors on the torque and flux-linkage of DTP-PMSM. If the duty cycles of the master–slave virtual vectors are $d_m$ and $d_s$, respectively, the torque change equation in the entire control period is shown as follows:

$$\begin{aligned} \Delta T_e &= \Delta T_{em} + \Delta T_{es} + \Delta T_{e0} \\ &= T_s(\lambda_{Tm}d_m + \lambda_{Ts}d_s - \lambda_e) \end{aligned} \tag{12}$$

$$\frac{\Delta T_e}{L_T} = \lambda_{Tm}d_m + \lambda_{Ts}d_s - \lambda_e \tag{13}$$

where $T_s$ is the control period; $\Delta T_e$ is the torque error; $L_T$ is the torque coefficient; and $\Delta T_{em}$, $\Delta T_{es}$, and $\Delta T_{e0}$ are the master–slave virtual vector torque changes and the zero vector torque changes, respectively. Similarly, the flux-linkage equation can be expressed as:

$$\frac{\Delta \psi_s}{L_\psi} = \lambda_{\psi m}d_m + \lambda_{\psi s}d_s \tag{14}$$

where $\Delta \psi_s$ is the flux-linkage error; $L_\psi$ is the flux-linkage coefficient; and $\lambda_{\psi m}$ and $\lambda_{\psi s}$ are the flux evaluation functions of master–slave virtual vectors, respectively. The master–slave virtual vector duty cycles dm and ds can be calculated by Equations (12) and (13). When the stator flux-linkage is located in sector I, the adjustment range of the master–slave virtual vectors when they act together is shown in Figure 5.

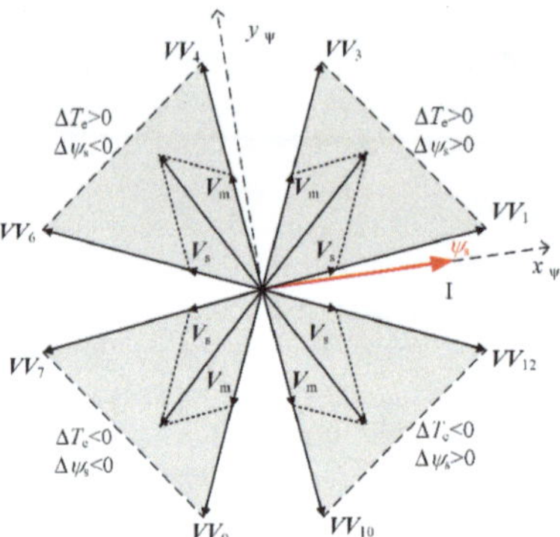



**Figure 5.** Adjustment range of master–slave virtual vectors.

Due to the large fluctuation of the torque and flux-linkage of DTP-PMSMs, and the imprecision of the torque coefficient and flux-linkage coefficient, the calculated value of the duty cycle of the two virtual vectors may not be between 0 and 1, so it is necessary to reassign the duty cycle of the master–slave virtual vectors. According to the principle of master virtual vector priority, it can be divided into the following distribution methods:

(1) When $d_m$ and ds are both less than zero, then the zero vector acts on the whole period;

(2) When $d_m$ is greater than zero or $d_s$ is less than zero, then the main virtual vector acts on the whole period;

(3) When the sum of $d_m$ and $d_s$ is less than 1, the master–slave virtual vectors and zero vector work together on a period;

(4) When $d_m$ is greater than zero and less than 1 and the sum of $d_m$ and $d_s$ is greater than 1, then the master–slave virtual vectors act on a period, where the duty cycle of the slave virtual vector is 1-$d_m$.

*4.3. Switching Sequence Correction*

The switch sequence generated by the synthesis of master–slave virtual vectors in the previous section cannot ensure symmetry, while the constantly changing angle value of $\theta_s$ leads to uncertainty in the duty cycles $d_m$ and $d_s$ of master–slave virtual vectors calculated using the evaluation function, further increasing the difficulty in generating a switch sequence. To minimize the disadvantages caused by asymmetric switching sequences, modifications can be made to the switching sequence generated via master–slave virtual vectors based on the output voltage duty cycle relationship for each phase. In the allocation modes mentioned in the previous section, modes (1) and (2) are single-vector actions that do not require modification to the switching sequence. In mode (3), uncertainty exists regarding the relationship between master–slave virtual vector duty cycles, making it difficult to correct the switching sequence. In mode (4), where the sum of the duty cycle for master–slave virtual vectors is 1, their switching sequences across different duty cycle ranges can be corrected through calculation.

Taking the first type of virtual vector acting on the motor in mode (4) as an example, when the stator flux-linkage is located in sector I, $VV_3$ is the master virtual vector and $VV_1$ is the slave virtual vector. The phase duty at this time can be calculated using Equation (7), as shown in Table 2. When the duty cycle of phase A and phase B is equal, $d_m$ is 0.73. When the duty cycle of phase V and phase W is equal, $d_m$ is 0.5. Furthermore, when $d_m$ is in different interval ranges, different correction methods can be adopted, as shown in Figure 6. When $0 < d_m \leq 0.27$, the effective vectors $V_{44}$ and $V_{65}$ are replaced by the intermediate vectors $V_{04}$, $V_{64}$, and $V_{67}$ to form a symmetric switching sequence. When $0.27 < d_m \leq 0.73$, the effective vectors $V_{24}$ and $V_{65}$ are replaced with the intermediate vectors $V_{04}$, $V_{64}$, and $V_{67}$. When $0.73 < d_m < 1$, the effective vectors $V_{24}$ and $V_{66}$ are replaced with the intermediate vectors $V_{04}$, $V_{64}$, and $V_{67}$. By analogy, the correction methods of switch sequences in different parity sectors can be obtained.

**Table 2.** Duty cycle of each phase for sectors I and II.

| Sector | I | II |
|---|---|---|
| | A $1 - 0.27d_{\mathrm{m}}$ | A $1 - 0.73d_{\mathrm{m}}$ |
| | B $0.27 + 0.73d_{\mathrm{m}}$ | B $0.73 + 0.27d_{\mathrm{m}}$ |
| Duty cycle of | C $0$ | C $0$ |
| each phase | U $1$ | U $1 - 0.27d_{\mathrm{m}}$ |
| | V $0.73d_{\mathrm{m}}$ | V $0.27 + 0.73d_{\mathrm{m}}$ |
| | W $0.27 - 0.27d_{\mathrm{m}}$ | W $0$ |
| Critical value of $d_{\mathrm{m}}$ | $0.27, 0.73$ | $0.5, 0.73$ |

The action time of the high level of each phase of the modified switching sequence does not change, and the modified synthetic vector does not change, which ensures that the effect of the master–slave virtual vector in the two subspaces remains unchanged. Only when the duty cycle of the master–slave virtual vector changes is the individual action vector replaced with the intermediate vector, and each parity sector can be modified to the symmetric waveform via the equivalent substitution of the intermediate vector within the above three value ranges of $d_{\mathrm{m}}$. Similarly, the switching sequence of the master–slave virtual vector in synthesis mode 2 can also be modified symmetrically according to this law. The modified master–slave virtual vector switch sequence is symmetrical in the center, which is convenient for hardware implementation, and also ensures that the power device operates once in one period, reducing the switching loss.

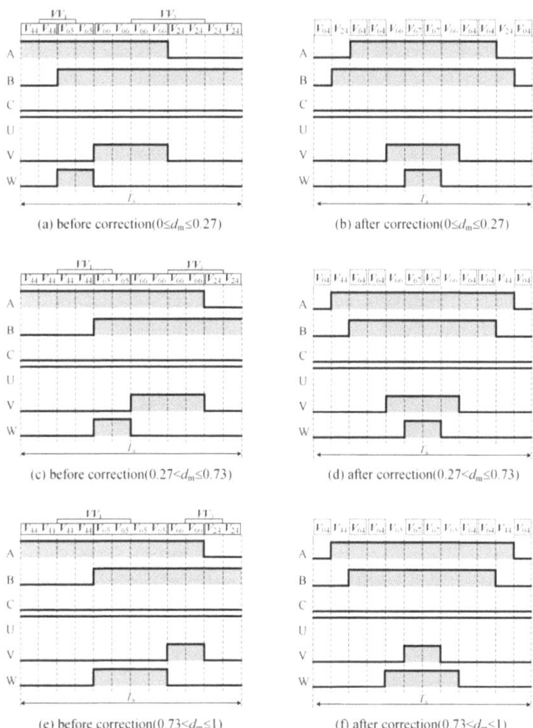

Figure 6. Schematic diagram of switch sequence before and after correction.

The overall block diagram of the master–slave virtual vector DTC of DTP-PMSMs based on duty cycle allocation is shown in Figure 7.

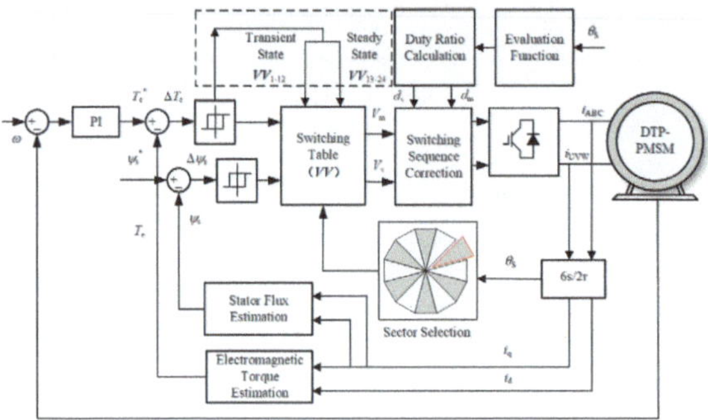

**Figure 7.** Direct torque control block diagram based on master–slave vector duty cycle allocation.

## 5. Experimental Verification

### 5.1. Experiment System

In order to verify the feasibility of the strategy proposed in this paper, a DTP-PMSM system experiment platform is built, as shown in Figure 8. The rated speed of the dual three-phase permanent magnet synchronous motor, which is manufactured by Hebei Electric Machinery Factory in China, used in the experiment is 1000 r/min, the rated torque is 10 Nm, the permanent magnet flux is 0.22 Wb, the pole-pairs number is 5, the d-axis inductance is 29 mH, and the q-axis inductance is 42 mH, respectively. The control algorithm is implemented with the TMS320F28377D DSP produced by TI, and Cyclone V FPGA produced by Intel. DSP is applied to execute the algorithm, and FPGA is applied to implement the high-precision analog-to-digital conversion (ADC) sampling, digital-to-analog conversion (DAC) conversion, and PWM pulse generation. The switching frequency of the IPM is 10 kHz.

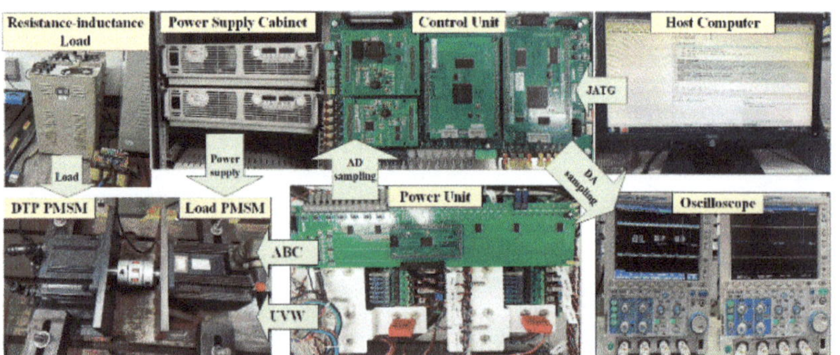


**Figure 8.** Experimental platform for the DTP-PMSM control system.

The experimental platform consists of a power loop in the strong electricity part and a control loop in the weak electricity part. After the experimental platform is powered on and debugged, Code Composer Studio (ccs10.0) is used to build the DSP experimental program. The experiment simulates the variable load of the motor through the series inductance of the resistance box. In the strong electricity part of the experimental platform, the DC bus voltage of the system is powered with the KEYSIGHT DC power supply. In the pulse transmission process of the system, the duty cycle signal is first sent from the main control panel to the CPLD panel. In the CPLD, the signal is converted into an optical signal through the optocoupler module, and then the 12-channel PWM wave is sent to the IPM through the optical fiber. The reference voltage is synthesized by controlling the opening and closing of the IGBT switch tube in the IPM module. The control loop of the weak electricity part of the experimental platform is composed of the DSP chip and the FPGA chip. The DSP chip mainly performs the operation of the basic algorithm of the motor. The DSP converts the algorithm into a duty cycle and controls the FPGA chip to generate a pulse signal. The pulse signal is transmitted to the CPLD board to generate a 12-channel PWM wave and is converted into the optical signal to control the opening and closing of the IPM.

### 5.2. Analysis of Steady-State Experimental Results

To verify the steady-state performance of the proposed strategy, the motor runs at 300 r/min with a load of 4 Nm. After steady-state operation, the experimental waveforms of the traditional strategy, the virtual vector strategy, and the proposed strategy are shown in Figure 9.

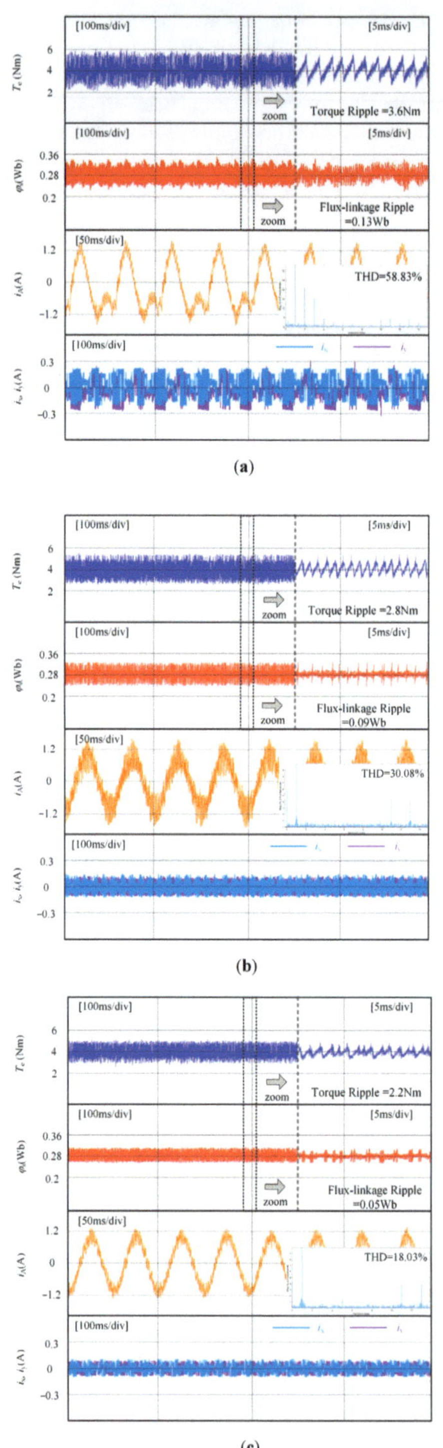


**Figure 9.** Steady-state operation experimental waveforms: (**a**) traditional strategy; (**b**) virtual vector strategy; and (**c**) proposed strategy.

By referring to the literature on DTC strategy in recent years, most of the improved DTC strategies are mainly analyzed from the three perspectives of the harmonic content analysis of motor current, torque ripple, and flux-linkage ripple. The traditional DTC strategy does not consider the voltage vector synthesis of the harmonic subspace, but the proposed control strategy can suppress the voltage synthesis of the harmonic subspace by redistributing the vector operation time. By analyzing the experimental results in Figure 9, under steady-state conditions, the phase current THD of the traditional DTC strategy is 58.8%, the torque ripple is 3.6 Nm, and the flux-linkage ripple is 0.13 Wb. However, the phase current THD of the proposed strategy is 18.0%, the torque ripple is 2.2 Nm, and the flux-linkage ripple is 0.05 Wb. Compared with the traditional DTC control strategy, the current THD of the proposed strategy is reduced by 69.4%, the torque ripple is reduced by 39%, and the flux-linkage ripple is reduced by 62%, which verifies the effectiveness of the proposed control strategy. Moreover, the experimental analysis results show that the proposed control strategy can effectively suppress the harmonic subspace current $i_x$, $i_y$. By analyzing the basic mathematical Equations (3)–(6) of the motor, it can be concluded that the harmonic subspace current will affect the flux-linkage of the harmonic subspace, which will adversely affect the flux-linkage $\varphi_s$ of the motor.

In addition, by analyzing other works in the literature [23], in order to further analyze the improvement effect of the proposed control strategy on motor flux-linkage and torque, two new variables are introduced: the torque standard deviation $\sigma_{Te}$ and the flux-linkage standard deviation $\sigma_{\varphi s}$. By introducing these two new evaluation variables, the effect of the proposed control strategy is further verified. After the analysis, it can be obtained that $\sigma_{Te}$ is reduced by 43% and $\sigma_{\varphi s}$ is reduced by 27% compared with the traditional control strategy under 300 r/min with a load of 4 Nm condition.

In order to further verify the steady-state performance of the proposed strategy under different conditions, the phase current THD, torque ripple values, and flux-linkage ripple values of the three control strategies under different load values are shown in Table 3 at the reference speed of 300 r/min.

**Table 3.** Steady-state performance comparison under different load conditions.

| | Control Strategy | 4 Nm | 6 Nm | 8 Nm |
|---|---|---|---|---|
| Current THD (%) | a | 58.8 | 42.7 | 36.5 |
| | b | 30.1 | 24.6 | 20.8 |
| | c | 18.0 | 16.2 | 14.8 |
| Torque Ripple (Nm) | a | 3.6 | 4.5 | 5.3 |
| | b | 2.8 | 3.7 | 4.2 |
| | c | 2.2 | 3.3 | 3.6 |
| Flux-linkage Ripple (Wb) | a | 0.13 | 0.14 | 0.16 |
| | b | 0.09 | 0.11 | 0.14 |
| | c | 0.05 | 0.06 | 0.08 |

From Table 3, it is obvious that compared to the traditional DTC strategy and virtual vector strategy, the strategy proposed in this paper reduces flux-linkage fluctuation and torque fluctuation under various conditions. After using two sets of virtual vectors, the phase current waveform is greatly improved.

*5.3. Analysis of Dynamic Experimental Results*

As shown in Figures 10 and 11, so as to verify the dynamic tracking performance of the proposed strategy, an experimental analysis was carried out under three different working conditions: the reference speed in the first working condition jumped from 300 r/min to 500 r/min and the second condition is that the motor is suddenly loaded by 4 Nm during no-load time until stable operation.

From Figure 10, it is obvious that the two control strategies can quickly track the reference value of the speed after the motor is started, there is no large overshoot phenomenon, and the speed of the proposed control strategy reaches the given reference value after a short adjustment when the speed changes at different speeds. It can be seen that the proposed strategy has good speed dynamic control performance.

As can be seen from Figure 11, both control strategies quickly reach the same torque value after the abrupt addition of a 4 Nm load. Therefore, the proposed strategy retains the advantages of the traditional strategy of fast response, and the proposed strategy can effectively suppress the torque fluctuation and flux-linkage fluctuation of DTP-PMSMs.

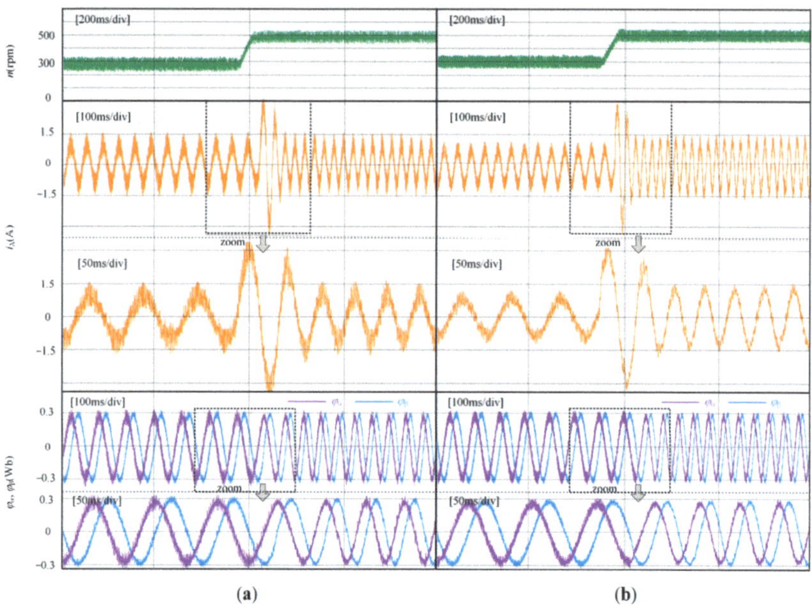

**Figure 10.** Experimental waveform of step speed response: (**a**) virtual vector strategy; (**b**) proposed strategy.

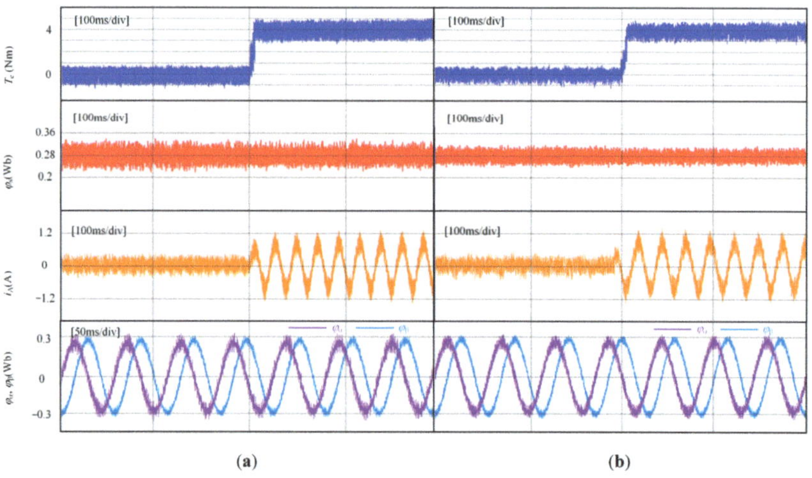

**Figure 11.** Load dynamic experimental waveform: (**a**) virtual vector strategy; (**b**) proposed strategy.

## 6. Conclusions

In this paper, a DTC strategy combining master–slave virtual vector duty cycle allocation is proposed for a DTP-PMSM system. Through experimental verification, the conclusions are as follows:

(1)  The use of virtual vectors reduces the voltage amplitude of the $x$-$y$ subspace to zero, effectively reducing the 5th and 7th harmonics, and also uses two types of virtual vectors to increase the torque control accuracy;

(2)  Using the exact value of the evaluation function to calculate and assign the duty cycle of the master–slave virtual vector, the torque fluctuation and flux-linkage fluctuation are reduced, and the adjustment range of the vector is increased;

(3)  The switching sequence of the master–slave virtual vector is remodified, and the effective vector action sequence under different sectors and different duty cycle values is analyzed, which reduces the difficulty of hardware implementation and ensures the effectiveness of harmonic suppression.

**Author Contributions:** Conceptualization, Q.G.; Data Curation, Z.Q.; Formal Analysis, G.Z.; Funding Acquisition, Z.Z. and Q.G.; Investigation, X.J.; Methodology, Z.Z.; Project Administration, Q.G.; Resources, Q.G.; Software, X.J.; Supervision, G.Z. and Q.G.; Validation, X.J.; Visualization, Z.Q.; Writing—Original Draft, Z.Q. All authors have read and agreed to the published version of the manuscript.

**Funding:** This research was supported in part by the National Natural Science Foundation of China under Grant 52077154, in part by the Natural Science Foundation of the Tianjin Municipality under Grant 23JCYBJC00290, and in part by the Zhejiang Provincial Natural Science Foundation of China under Grant LY24E070003.

**Institutional Review Board Statement:** Not applicable.

**Informed Consent Statement:** Not applicable.

**Data Availability Statement:** The original contributions presented in the study are included in the article, further inquiries can be directed to the corresponding author.

**Conflicts of Interest:** The authors declare no conflicts of interest.

# References

1.  Hu, Y.; Zhu, Z.; Odavic, M. Comparison of Two-Individual Current Control and Vector Space Decomposition Control for Dual Three-Phase PMSM. *IEEE Trans. Ind. Appl.* **2017**, *53*, 4483–4492. [CrossRef]
2.  Kallio, S.; Andriollo, M.; Tortella, A.; Karttunen, J. Decoupled d-q Model of Double-Star Interior-Permanent-Magnet Synchronous Machines. *IEEE Trans. Ind. Electron.* **2013**, *60*, 2486–2494. [CrossRef]
3.  Huang, L.; Ji, J.; Zhao, W.; Tang, H.; Tao, T.; Jin, S. Direct torque control for dual three-phase permanent magnet motor with improved torque and flux. *IEEE Trans. Energy Convers.* **2022**, *37*, 2385–2397. [CrossRef]
4.  Zhao, Y.; Lipo, T.A. Space vector PWM control of dual three-phase induction machine using vector space decomposition. *IEEE Trans Ind. Appl.* **1995**, *31*, 1100–1109. [CrossRef]
5.  Demir, Y.; Aydin, M. A Novel Dual Three-Phase Permanent Magnet Synchronous Motor with Asymmetric Stator Winding. *IEEE Trans. Magn.* **2016**, *52*, 1–5. [CrossRef]
6.  Huang, L.; Ji, J.; Zhao, W.; Tao, T.; Cui, J. Duty-ratio-based direct torque control with enhanced harmonic current suppression for dual-three-phase permanent magnet motor. *IEEE Trans. Power Electron.* **2022**, *37*, 11098–11108. [CrossRef]
7.  Gonzalez-Prieto, I.; Duran, M.J.; Aciego, J.J.; Martin, C.; Barrero, F. Model Predictive Control of Six-Phase Induction Motor Drives Using Virtual Voltage Vectors. *IEEE Trans. Ind. Electron.* **2018**, *65*, 27–37. [CrossRef]
8.  Xu, J.; Odavic, M.; Zhu, Z.; Wu, Z.; Nuno, F. Switching-table-based direct torque control of dual three-phase PMSMs with closed-loop current harmonics compensation. *IEEE Trans. Power Electron.* **2021**, *36*, 10645–10659. [CrossRef]
9.  Ren, Y.; Zhu, Z. Enhancement of Steady-State Performance in Direct-Torque-Controlled Dual Three-Phase Permanent-Magnet Synchronous Machine Drives with Modified Switching Table. *IEEE Trans. Ind. Electron.* **2015**, *62*, 3338–3350. [CrossRef]
10.  Shao, B.; Zhu, Z.; Feng, J.; Guo, S.; Li, Y.; Liao, W. Compensation of Selective Current Harmonics for Switching-Table-Based Direct Torque Control of Dual Three-Phase PMSM Drives. *IEEE Trans. Ind. Appl.* **2021**, *57*, 2505–2515. [CrossRef]
11.  Ren, Y.; Zhu, Z.; Green, J.E.; Li, Y.; Zhu, S.; Li, Z. Improved Duty-Ratio-Based Direct Torque Control for Dual Three-Phase Permanent Magnet Synchronous Machine Drives. *IEEE Trans. Ind. Appl.* **2019**, *55*, 5843–5853. [CrossRef]

12. Ren, Y.; Zhu, Z. Reduction of Both Harmonic Current and Torque Ripple for Dual Three-Phase Permanent-Magnet Synchronous Machine Using Modified Switching-Table-Based Direct Torque Control. *IEEE Trans. Ind. Electron.* **2015**, *62*, 6671–6683. [CrossRef]
13. Li, L.; Zhu, G. Electromagnetic-thermal-stress efforts of stator-casing grease buffers for permanent magnet driving motors. *IEEE Trans. Ind. Appl.* **2023**, *60*, 1268–1276. [CrossRef]
14. Wang, X.; Wang, Z.; Xu, Z. A Hybrid Direct Torque Control Scheme for Dual Three-Phase PMSM Drives with Improved Operation Performance. *IEEE Trans. Power Electron.* **2019**, *34*, 1622–1634. [CrossRef]
15. Luo, Y.; Liu, C. Model Predictive Control for a Six-Phase PMSM With High Robustness Against Weighting Factor Variation. *IEEE Trans. Ind. Appl.* **2019**, *55*, 2781–2791. [CrossRef]
16. Luo, Y.; Liu, C. A Flux Constrained Predictive Control for a Six-Phase PMSM Motor with Lower Complexity. *IEEE Trans. Ind. Electron.* **2019**, *66*, 5081–5093. [CrossRef]
17. Luo, Y.; Liu, C. A Simplified Model Predictive Control for a Dual Three-Phase PMSM With Reduced Harmonic Currents. *IEEE Trans. Ind. Electron.* **2018**, *65*, 9079–9089. [CrossRef]
18. Deng, W.; Zhang, X.; Yan, B. An Enhanced Discrete Virtual Vector-Based Direct Torque Control of PMSM Drives. *IEEE Trans. Energy Convers.* **2024**, *39*, 277–286. [CrossRef]
19. Liu, Y.; Lyu, M.; Huang, S.; Liao, W.; Huang, S. Direct Torque Control Schemes for Dual Three-Phase PMSM Considering Unbalanced DC-link Voltages. *IEEE Trans. Energy Convers.* **2024**, *39*, 229–242. [CrossRef]
20. Qiu, X.; Ji, J.; Zhou, D.; Zhao, W.; Chen, Y.; Huang, L. A modified flux observer for sensorless direct torque control of dual three-phase PMSM considering open-circuit fault. *IEEE Trans. Power Electron.* **2022**, *37*, 15356–15369. [CrossRef]
21. Xia, C.; Zhao, J.; Yan, Y.; Shi, T. A Novel Direct Torque and Flux Control Method of Matrix Converter-Fed PMSM Drives. *IEEE Trans. Power Electron.* **2014**, *29*, 5417–5430. [CrossRef]
22. Xia, C.; Zhao, J.; Yan, Y.; Shi, T. A Novel Direct Torque Control of Matrix Converter-Fed PMSM Drives Using Duty Cycle Control for Torque Ripple Reduction. *IEEE Trans. Ind. Electron.* **2014**, *61*, 2700–2713. [CrossRef]
23. Shao, B.; Zhu, Z.; Yan, L.; Feng, J.; Guo, S.; Li, Y.; Ling, F. Torque ripple reduction for direct torque control of dual three-phase PMSM based on multiple virtual voltage vectors. *IEEE Trans. Energy Convers.* **2022**, *38*, 296–309. [CrossRef]

*World Electric Vehicle Journal*

*Article*

# Online Inductance Identification of Permanent Magnet Synchronous Motors Independent of Rotor Position Information

**Jilei Xing [1],*, Junzhi Zhang [1], Xingming Zhuang [2] and Yao Xu [3]**

[1]  School of Vehicle and Mobility, Tsinghua University, Beijing 100084, China; powertrain@tsinghua.edu.cn
[2]  BIT HuaChuang Electric Vehicle Technology Co., Ltd., Beijing 100081, China; zhuangxingming@huachuangev.com
[3]  School of Mechanical Engineering, Beijing Institute of Technology, Beijing 100081, China; xuyao@bit.edu.cn
*  Correspondence: xingjilei699@mail.tsinghua.edu.cn; Tel.: +86-18813009025

**Abstract:** Sensorless control of permanent magnet synchronous motors is preferable in some applications due to cost and mounting space concerns. The performance of most existing position estimation methods greatly depends on the accuracy of the motor inductance. As the estimated position should not be involved in the parameter identification process in a sensorless control system, an online inductance identification method independent of the rotor position information is developed in this paper. The proposed method utilizes the recursive least square algorithm and the particle swarm optimization algorithm to realize real-time identification of the inductance along the direct axis and the quadrature axis, respectively, based on the deduced parametric equations without position information. The proposed method is efficient enough to be implemented within 0.2 ms and does not introduce any additional signal injection. A test bench is built to validate the characteristics of the method, and the experimental results show that the identified inductance can converge to the actual value rapidly and is robust to changes in the initial values and stator current. With the proposed method, accurate estimation of the rotor position and speed can be obtained using traditional model-based position estimators, and the stability of the sensorless control system can be improved significantly.

**Keywords:** permanent magnet synchronous motors; online inductance identification; sensorless control; recursive least square algorithm; particle swarm optimization

**Citation:** Xing, J.; Zhang, J.; Zhuang, X.; Xu, Y. Online Inductance Identification of Permanent Magnet Synchronous Motors Independent of Rotor Position Information. *World Electr. Veh. J.* **2024**, *15*, 35. https://doi.org/10.3390/wevj15010035

Academic Editor: Ziqiang Zhu

Received: 15 December 2023
Revised: 12 January 2024
Accepted: 15 January 2024
Published: 22 January 2024



## 1. Introduction

Permanent magnet synchronous motors (PMSMs) have been extensively adopted in electric vehicles in recent years due to positive features such as high efficiency, high reliability and high power density [1]. In most applications, the rotor position of the PMSM is obtained using a position sensor for precise control of the motor. However, the position sensor is eliminated in consideration of the cost, mounting space and operating environment in applications like power steering pumps and compressors in electric commercial vehicles [2,3]. And various rotor position estimation methods are substituted for position sensors under these circumstances. Equally, a sensorless control system can be used for fault-tolerant control of the traction motors in electric vehicles.

The existing rotor position estimation methods can be classified into two categories: signal injection methods and model-based methods [4]. Signal injection methods take advantage of the salient effect of PMSMs. The rotor position information can be extracted from the current response after injecting specific high-frequency voltage into the stator windings [5,6]. These methods are effective at standstill and low speeds but give bad performance as the speed increases [7]. Moreover, signal injection methods are not suitable for surface-mounted permanent magnet synchronous motors or negative-saliency permanent magnet synchronous motors. Model-based rotor position estimation methods can be further divided into back electromotive force (EMF)-based methods and active-flux-based methods [8,9]. These methods utilize a mathematical model of the PMSM to construct different kinds of position observers, such as the sliding-mode observer [10], the extended Kalman filter observer [11] and the adaptive observer [12]. The model-based methods can give precise estimation results at nominal and high speeds. And the active flux-based methods have been proved to detect the accurate rotor position at low speeds [13]. It is obvious that model-based methods are indispensable in sensorless control systems of PMSMs, and the position estimation precision greatly depends on the accuracy of the model parameters [14].

Generally, the model parameters of PMSMs refer to the winding phase resistance, the permanent magnet flux linkage and the inductance along the direct axis (d-axis) and the quadrature axis (q-axis). Though the nominal values of these four parameters are available from manufacturers or using offline parameter measurement, these parameters vary with operating environment and the conditions of the PMSMs. More specifically, the winding phase resistance and the permanent magnet flux linkage are mainly affected by the ambient temperature [15]. And variation in the inductance is attributed to the magnetic saturation effect, which is mainly caused by changes in stator current [16]. Apparently, stator current varies much more rapidly than temperature in practice, resulting in more frequent changes in inductance than in the other two parameters. Furthermore, research on parameter mismatching effects on position estimation methods indicates that the inaccuracy of the inductance is the dominant factor in position estimation errors [17]. It is suggested that an inductance lookup table should be mapped offline and stored for reliable control in applications [18]. However, offline calibration is time-consuming and difficult to implement in some situations. As a result, online inductance identification is of great importance for improving the performance of sensorless control systems.

The online inductance identification methods of PMSMs have been intensively studied in recent decades, and the related solutions include numerical methods, observer-based methods and AI-based methods [19]. Most of these methods are used for control systems with position sensors, and the sampled position information is indispensable in the identification process. In recent years, online inductance identification in sensorless control systems has attained more discussion. The traditional solution in sensorless control systems is calculating the current and voltage components in the rotating reference frame using the estimated rotor position instead of the actual rotor position. Then, the current and voltage components can be applied in the identification process [20]. Under this theory, researchers pay more attention to the problem of parameter mismatching but ignore the adverse effect on the parameter identification accuracy caused by position estimation errors. In order to realize online inductance identification independent of position information, several methods have been proposed. One effective method is obtaining the inductance from the current response under high-frequency voltage injection [21]. But the extra excitation voltage may cause torque ripples and power loss, and even not be allowed in some operating conditions [22]. The recursive least square (RLS) method is employed to identify the inductance and resistance simultaneously in [23]. A set of complicated state equations is established, and 10 intermediate variables are identified using RLS to calculate the motor parameters. Though the rotor position is not used, the method may give ill-converged results because of the rank-deficient problem. A multi-parameter identification method is proposed in [24] based on the affine projection algorithm (APA) and multiscale framework. The rank-deficient problem is solved by estimating different parameters in different timescales. However, components of the EMF in the estimated reference frame are used to develop the APA, which contains the position information. In [25], a set of polynomial equations is used to compute the initial values of the motor parameters and the rotor position numerically, while the availability in rotating conditions is not verified. Another solution for online inductance identification in a sensorless control system is state observers. For instance, the sliding-mode observer proposed in [26] can estimate the inductance and rotor position simultaneously. This method is better suited to surface-mounted PMSMs, which have equal inductance along the d-axis and q-axis, because only one parameter is treated as unknown in the observer while the others are taken as constant. Research on applications of AI-based methods in sensorless control systems is reported in [27,28]: these methods can give accurate identification results in different operating conditions but suffer from the calculation burden and a long training process.

In this paper, a novel online inductance identification method independent of the rotor position information is introduced to estimate the inductance along the d-axis and q-axis simultaneously without any extra excitation or rank-deficient problem. Compared with previous studies, the paper provides more detailed analysis of the adverse effect caused by position estimation errors in online inductance identification, drawing the conclusion that the estimated position should not be involved in the parameter identification process of PMSMs. Mathematical models of the PMSM in a stationary reference frame and rotating reference frame are utilized together to deduce parametric equations of inductance which do not contain the rotor position information. Then, the deduced equations are combined with RLS and particle swarm optimization (PSO) to give precise estimation of the inductance. With the proposed method, the estimation results on the inductance will not be affected by the accuracy of the rotor position and can converge to the accurate value rapidly, which will further improve the performance of the sensorless control system. It has to be mentioned that the proposed online inductance identification method is suitable for both surface PMSMs and interior PMSMs and can be applied along with various kinds of model-based position estimators. The proposed method can also benefit field-oriented control and fault diagnosis in various applications of PMSMs. For instance, the proposed method can be used to construct a reliable sensorless control system for traction motors in electric vehicles in case of failures of the position sensors. The effectiveness and feasibility are validated using simulations and experiments.

The rest of the paper is organized as follows. The mathematical model and sensorless control system of the PMSM are introduced in Section 2. The effect on the identification accuracy caused by position estimation errors is analyzed in Section 3. Then, the online inductance identification method independent of the rotor position is proposed in Section 4. The performance of the proposed identification method is verified using the experimental results in Section 5. Finally, the key features of the online inductance identification method are summarized in Section 6.

## 2. Sensorless Control Systems

### 2.1. Mathematical Model of PMSMs

Model-based rotor position estimation methods are usually derived from the voltage equations of PMSMs, which can be written as follows in the rotating reference frame [29].

$$\begin{bmatrix} u_d \\ u_q \end{bmatrix} = \begin{bmatrix} R_s + pL_d & 0 \\ 0 & R_s + pL_d \end{bmatrix} \begin{bmatrix} i_d \\ i_q \end{bmatrix} + \omega_e \begin{bmatrix} -\psi_q \\ \psi_d \end{bmatrix} \tag{1}$$

where $u_d, u_q, i_d, i_q, L_d, L_q, \psi_d$ and $\psi_q$ are the stator voltage components, stator current components, inductance and flux linkage along the d-axis and the q-axis, $\omega_e$ is the electrical speed, $R_s$ is the phase winding resistance and $p$ is the differential operator. The stator flux linkage can be expressed as:

$$\begin{bmatrix} \psi_d \\ \psi_q \end{bmatrix} = \begin{bmatrix} L_d & 0 \\ 0 & L_q \end{bmatrix} \begin{bmatrix} i_d \\ i_q \end{bmatrix} + \begin{bmatrix} \psi_f \\ 0 \end{bmatrix} \tag{2}$$

where $\psi_f$ is the permanent magnet flux linkage. In order to simplify the model of the PMSM in the stationary frame, the extended back EMF is introduced as:

$$E_{ext} = (L_d - L_q)\omega_e i_d + (L_q - L_d)pi_q + \omega_e \psi_f \tag{3}$$

Then, (1) can be rewritten as:

$$\begin{bmatrix} u_d \\ u_q \end{bmatrix} = \begin{bmatrix} R_s + pL_d & -\omega_e L_q \\ \omega_e L_q & R_s + pL_d \end{bmatrix} \begin{bmatrix} i_d \\ i_q \end{bmatrix} + \begin{bmatrix} 0 \\ E_{ext} \end{bmatrix} \tag{4}$$

And (4) can be converted into the stationary reference frame as:

$$\begin{bmatrix} u_\alpha \\ u_\beta \end{bmatrix} = \begin{bmatrix} R_s + pL_d & -\omega_e \Delta L \\ \omega_e \Delta L & R_s + pL_d \end{bmatrix} \begin{bmatrix} i_\alpha \\ i_\beta \end{bmatrix} + \begin{bmatrix} e_{ext,\alpha} \\ e_{ext,\beta} \end{bmatrix} \tag{5}$$

where $u_\alpha$, $u_\beta$, $i_\alpha$ and $i_\beta$ are the stator voltage components and stator current components in the stationary $\alpha$-$\beta$ reference frame, and $\Delta L$ is the difference between $L_q$ and $L_d$. $e_{ext,\alpha}$ and $e_{ext,\beta}$ are the $\alpha$-axis and $\beta$-axis components of $E_{ext}$, which can be expressed as:

$$\begin{bmatrix} e_{ext,\alpha} \\ e_{ext,\beta} \end{bmatrix} = E_{ext} \begin{bmatrix} -\sin\theta_e \\ \cos\theta_e \end{bmatrix} \tag{6}$$

where $\theta_e$ is the electrical rotor position.

Similarly, the concept of active flux is utilized to present the mathematical model in the stationary frame in a simple form. The active flux of the PMSM is defined as [30]:

$$\psi_{ext} = \psi_f - (L_q - L_d) i_d \tag{7}$$

By combining (6) and (1), the model of the PMSM can be rewritten as:

$$\begin{bmatrix} u_d \\ u_q \end{bmatrix} = \begin{bmatrix} R_s + pL_q & -\omega_e L_q \\ \omega_e L_q & R_s + pL_q \end{bmatrix} \begin{bmatrix} i_d \\ i_q \end{bmatrix} + \begin{bmatrix} p\psi_{ext} \\ \omega_e \psi_{ext} \end{bmatrix} \tag{8}$$

Furthermore, the $\alpha$-axis and $\beta$-axis components of the active flux are given as:

$$\begin{bmatrix} \psi_{ext,\alpha} \\ \psi_{ext,\beta} \end{bmatrix} = \psi_{ext} \begin{bmatrix} \cos\theta_e \\ \sin\theta_e \end{bmatrix} \tag{9}$$

The stator linkage in the stationary reference frame can be obtained by associating (9) with (2), as in (10).

$$\begin{bmatrix} \psi_\alpha \\ \psi_\beta \end{bmatrix} = \begin{bmatrix} L_q & 0 \\ 0 & L_q \end{bmatrix} \begin{bmatrix} i_\alpha \\ i_\beta \end{bmatrix} + \begin{bmatrix} \psi_{ext,\alpha} \\ \psi_{ext,\beta} \end{bmatrix} \tag{10}$$

And (8) can be transformed into the stationary reference frame with the help of (10).

$$\begin{bmatrix} u_\alpha \\ u_\beta \end{bmatrix} = \begin{bmatrix} R_s & 0 \\ 0 & R_s \end{bmatrix} \begin{bmatrix} i_\alpha \\ i_\beta \end{bmatrix} + p \begin{bmatrix} \psi_\alpha \\ \psi_\beta \end{bmatrix} \tag{11}$$

Figure 1 gives a summary of the equations above, including the extended back EMF model and the active flux model. $T_{park}$ in the diagram is the Park transformation matrix.

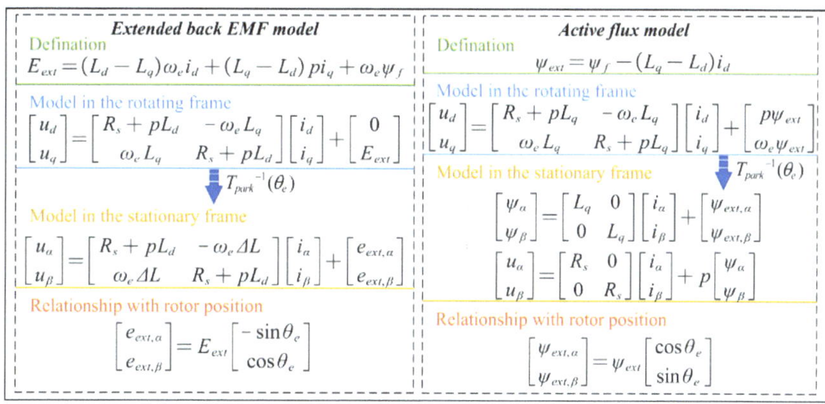

**Figure 1.** Illustration diagram of the model equations used in this paper.

*2.2. Rotor Position Estimator*

Both the extended back EMF and the active flux in the stationary reference frame contain the information on the electrical rotor position. Equations (5) and (11) can be used to construct the back EMF observer and the flux observer, respectively. It has to be noted that the performance of the proposed online inductance identification method will not be affected by the types of used models or observers, so a simple back-EMF-based position estimator is employed in this paper [31]. The extended back EMF observer can be expressed as:

$$p\begin{bmatrix} \hat{i}_\alpha \\ \hat{i}_\beta \end{bmatrix} = \frac{1}{L_d}\left\{ \begin{bmatrix} u_\alpha \\ u_\beta \end{bmatrix} - \begin{bmatrix} R_s & -\omega_e\Delta L \\ \omega_e\Delta L & R_s \end{bmatrix}\begin{bmatrix} \hat{i}_\alpha \\ \hat{i}_\beta \end{bmatrix} + \begin{bmatrix} \hat{e}_{ext,\alpha} \\ \hat{e}_{ext,\beta} \end{bmatrix} \right\} \tag{12}$$

where $\hat{i}_\alpha$, $\hat{i}_\beta$, $\hat{e}_{ext,\alpha}$ and $\hat{e}_{ext,\beta}$ are the estimated quantities of $i_\alpha$, $i_\beta$, $e_{ext,\alpha}$ and $e_{ext,\beta}$. The estimated electrical speed is regarded the actual speed, with the assumption that the error between them is sufficiently small. As the stator current components in the stationary reference frame are measured in real time using current sensors, the estimation error of $i_\alpha$ and $i_\beta$ can be calculated and used as the input to a PI compensator. The PI compensator is regarded to be the substitute for the unknown extended back EMF in (12), as shown in (13).

$$\begin{bmatrix} \hat{e}_{ext,\alpha} \\ \hat{e}_{ext,\beta} \end{bmatrix} = \left( K_p + \frac{K_i}{s} \right)\begin{bmatrix} \hat{i}_\alpha - i_\alpha \\ \hat{i}_\beta - i_\beta \end{bmatrix} \tag{13}$$

where $K_p$ and $K_i$ are the gains of the PI compensator. With the estimated extended back EMF, the rotor position can be estimated using the arctangent function and a phase-locked loop (PLL) observer. The structure of this rotor position estimator is given in Figure 2. The estimation accuracy is affected by multiple factors, including but not limited to calculation delay, parameter mismatching, improper gains, etc. The precision of the speed estimation is assured in most position estimators, but position estimation errors seem to be inevitable under different operating environments and conditions. Though the position estimation error will not cause the sensorless control system to malfunction, the accuracy of the traditional online inductance identification will be influenced greatly, which is discussed in detail in Section 3.

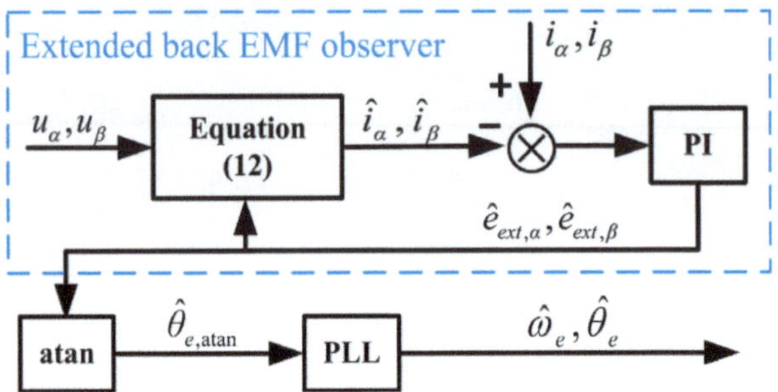



**Figure 2.** Structure of the extended back-EMF-based rotor position estimator.

### 3. Inductance Identification Accuracy Analysis with Position Error

The RLS method with a forgetting factor is widely adopted in the research on online motor parameter identification for its simplicity [32]. The typical system in the RLS can be expressed as:

$$Y(k) = W(k)^T X(k) \tag{14}$$

where $Y$ and $X$ are the output and input of the system, $W$ is the estimated parameter matrix and $k$ represents the number of iterations. The estimated parameter matrix can be obtained recursively as:

$$\begin{cases} \hat{W}(k) = \hat{W}(k-1) + K(k)\left[Y(k) - \hat{W}(k-1)^T X(k)\right] \\[2mm] K(k) = \dfrac{P(k-1)X(k)}{\lambda + X(k)^T P(k-1)X(k)} \\[2mm] P(k) = \dfrac{1}{\lambda}\left[I - K(k)X(k)^T\right]P(k-1) \end{cases} \tag{15}$$

The voltage equations in (1) and (2) are commonly utilized to construct the estimated system. The time-derivative terms can be neglected in the steady state, and the identification systems for inductances are given as:

$$\begin{cases} Y = u_d - R_s i_d \\ X = -\omega_e i_q \\ W = L_q \end{cases}, \quad \begin{cases} Y = u_q - R_s i_q - \omega_e \psi_f \\ X = \omega_e i_d \\ W = L_d \end{cases} \tag{16}$$

The above utilization of RLS is simple but effective in control systems of PMSMs with position sensors. However, the current and voltage components in the rotating reference frame will deviate from the actual values due to position estimation errors in sensorless control systems. Simulations of the above RLS identification method are executed to analyze its effect on online inductance identification. The motor parameters used in the simulations are given in Table 1. The initial values of the inductance are set to zero, and the motor operates under the rated conditions during the identification period. The online identification process is repeated, respectively, with the actual rotor position and position errors of −0.05 rad, 0.05 rad and 0.1 rad. The simulation results are shown in Figure 3a,b. It can be seen that all of the eight identification results converge to a stable value rapidly, but unbiased estimation can be realized only with the actual rotor position. The simulation results in Figure 3a show that the average relative identification error of $L_d$ is 0.62% compared to the actual rotor position, and increases to 11.29% with an error of −0.05 rad. And a positive position error will lead to negative relative errors, which reach −10.55% with an error of 0.05 rad and −18.04% with an error of 0.1 rad. A similar tendency can also be observed for the identification of $L_q$, as shown in Figure 3b.

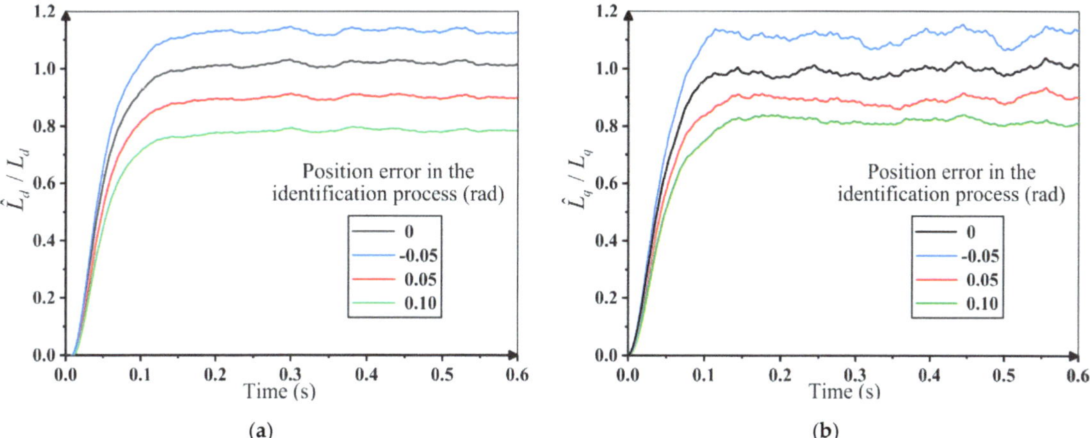

**Figure 3.** Inductance identification results using RLS with different rotor position errors: (**a**) Identification results of d-axis inductance; (**b**) identification results of q-axis inductance.

**Table 1.** Motor parameters.

| Denotation | Value |
| --- | --- |
| Rated power (kW) | 30 |
| Rated torque (N·m) | 100 |
| Rated speed (rpm) | 3000 |
| Number of pole pairs | 4 |
| Stator resistance ($\Omega$) | 0.02 |
| d-axis inductance (mH) | 0.3 |
| q-axis inductance (mH) | 0.6 |
| Flux linkage (Wb) | 0.081 |

## 4. Proposed Online Inductance Identification Method

### 4.1. Parametric Equation Derivation of q-Axis Inductance

The most direct approach to eliminating the effect of rotor position errors is finding parametric equations of the inductance which are independent of the position information. In consideration of the change rate and influence factors of the parameter variation, the resistance, the permanent magnet flux linkage and the rotor speed are thought to be of known quantities in the proposed inductance identification method. The estimated electrical speed is regarded the actual rotor speed. The resistance and the permanent magnet flux linkage can be obtained using offline measurement or corresponding online identification methods.

The stator current components can be given as follows in steady operating conditions.

$$\begin{bmatrix} i_\alpha \\ i_\beta \end{bmatrix} = I_s \begin{bmatrix} \cos \theta_i \\ \sin \theta_i \end{bmatrix} \tag{17}$$

where $I_s$ is the magnitude of the current vector and $\theta_i$ is the angle between the current vector and $\alpha$-axis. Then, the stator linkages in the stationary reference frame can be derived by combining (9), (10) and (17).

$$\begin{bmatrix} \psi_\alpha \\ \psi_\beta \end{bmatrix} = L_q I_s \begin{bmatrix} \cos \theta_i \\ \sin \theta_i \end{bmatrix} + \psi_{ext} \begin{bmatrix} \cos \theta_e \\ \sin \theta_e \end{bmatrix} \tag{18}$$

As the angle between the current vector and d-axis can be considered constant in every identification process, $\theta_i$ and $\theta_e$ share the same angular velocity. Taking the derivative of (18), the following equation can be developed:

$$p\begin{bmatrix} \psi_\alpha \\ \psi_\beta \end{bmatrix} = \omega_e\left\{ L_q I_s \begin{bmatrix} -\sin\theta_i \\ \cos\theta_i \end{bmatrix} + \psi_{ext}\begin{bmatrix} -\sin\theta_e \\ \cos\theta_e \end{bmatrix}\right\} = \omega_e\begin{bmatrix} -\psi_\beta \\ \psi_\alpha \end{bmatrix} \tag{19}$$

Then, expressions of the stator linkage are obtained by substituting (19) into (11).

$$\begin{bmatrix} \psi_\alpha \\ \psi_\beta \end{bmatrix} = \frac{1}{\omega_e}\begin{bmatrix} u_\beta \\ -u_\alpha \end{bmatrix} + \frac{R_s}{\omega_e}\begin{bmatrix} -i_\beta \\ i_\alpha \end{bmatrix} \tag{20}$$

Meanwhile, (10) can be rewritten as:

$$\psi_{ext}\begin{bmatrix} \cos\theta_e \\ \sin\theta_e \end{bmatrix} = \begin{bmatrix} \psi_\alpha \\ \psi_\beta \end{bmatrix} - \begin{bmatrix} L_q & 0 \\ 0 & L_q \end{bmatrix}\begin{bmatrix} i_\alpha \\ i_\beta \end{bmatrix} \tag{21}$$

It is obvious that $L_d$ is not involved in (21). And the rotor position can be removed by calculating the quadratic sum.

$$\psi_{ext}^2 = \left(\psi_\alpha - L_q i_\alpha\right)^2 + \left(\psi_\beta - L_q i_\beta\right)^2 \tag{22}$$

Considering the algebraic property of trigonometric functions, the estimated position error will not affect the accuracy of the amplitude of the active flux linkage. It is reasonable to bring the estimated amplitude of the active flux into the identification process without worrying about the effect of position errors. The estimated amplitude is directly available in active-flux-based position estimation methods. And in the extended back-EMF-based methods, the amplitude can be acquired approximately in steady conditions using:

$$\hat{\psi}_{ext} \approx \frac{\sqrt{\hat{e}_{ext,\alpha}^2 + \hat{e}_{ext,\beta}^2}}{\hat{\omega}_e} \tag{23}$$

With the help of (20) and (22), a parametric equation of $L_q$ can be derived as follows.

$$\psi_{ext}^2 - \left(\psi_\alpha^2 + \psi_\beta^2\right) = I_s^2 L_q^2 - 2\left(\psi_\alpha i_\alpha + \psi_\beta i_\beta\right)L_q \tag{24}$$

where $L_q$ is the only unknown, and the other quantities can be obtained independently, as mentioned above.

### 4.2. Online Identification of q-Axis Inductance Using PSO

It is evident that the parametric equation in (24) is a quadratic equation of $L_q$, which is not applicable in RLS identification methods. In this instance, computational-intelligence-based algorithms are widely adopted because they do not have special requirements in terms of the model structures. Among the general intelligent algorithms, PSO is noted for its simplicity and computational efficiency [33]. However, several iterations of the system model simulations are required in PSO to evaluate the candidate solutions and obtain the satisfactory one. So, most of the existing PSO-based identification methods are implemented offline [34]. Only if the simulations can be finished faster than the execution frequency of identification can online parameter identification using PSO be realized. Fortunately, the model of $L_q$ in (24) is quite simple. And with proper time-saving measures, it is possible to accomplish single-parameter identification in real time.

The PSO algorithm is inspired by the social intelligent behaviour of bird flocks. Assuming an m-dimensional solution space, each particle can be described by the location vector $X_i = (x_{i1}, x_{i2}, \ldots, x_{im})$ and the velocity vector $V_i = (v_{i1}, v_{i2}, \ldots, v_{im})$, where $i$ is the sequence number, $1 \leq i \leq I$ and $I$ is the sum of particles. In the k-th iteration, the locations of the particles are used to calculate fitness. According to the value of fitness, the best location of $X_i$ is marked as $Pbest_i(k)$, which is also called the particle best location. And the best location in the swarm is marked as $Gbest(k)$, which is also called the global best location. Then, the searching procedure can be executed by updating the velocity and the location as follows.

$$V_i(k+1) = \omega(k+1)V_i(k) + c_1 r_1 [Pbest_i(k) - X_i(k)] + c_2 r_2 [Gbest(k) - X_i(k)] \tag{25}$$

$$X_i(k+1) = X_i(k) + V_i(k+1) \tag{26}$$

where $c_1$ is the cognitive acceleration constant and $c_2$ is the social acceleration constant. $c_1$ and $c_2$ are suggested to be set within $[0,2]$, and $c_1$ should be less than $c_2$. $r_1$ and $r_2$ are two random numbers, and $\omega$ is the inertia weight factor, which is a positive number less than 1 and decreasing with iterations. The process of the PSO-based parameter identification method can be illustrated as follows, with Figure 4 as a reference.

Step 1: Design the fitness function according to the system model and collect the corresponding sample data.

Step 2: Initialize the swarm randomly and calculate the initial fitness to obtain the initial particle best location and global best location.

Step 3: Update $V_i$, $X_i$, $Pbest_i$ and $Gbest$ in multiple iterations until the terminate condition is satisfied. The terminate condition can be the number of iterations or the value of fitness.

Step 4: Output the value of $Gbest$ as the identification result.

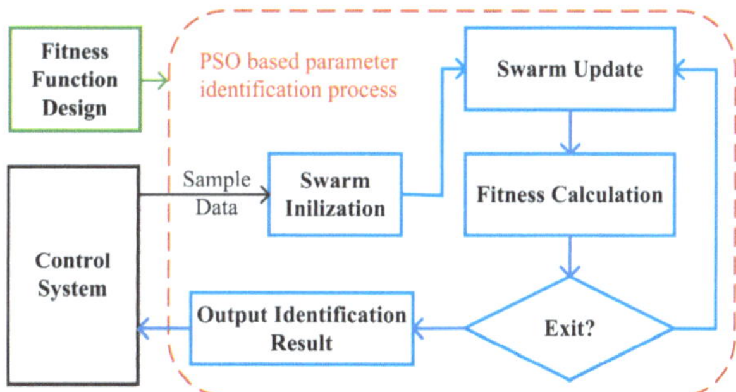


**Figure 4.** Block diagram of the PSO (particle swarm optimization)-based parameter identification method.

In the case of online inductance identification, the above procedures must be finished in several milliseconds or less to track the rapid variations. In consideration of the computing power of the hardware and the change rate of the inductance, the objective update frequency for the inductance in this paper is 1 kHz. The identification process will be implemented with the sampling process in parallel to the microprocessor as shown in Figure 5. It is obviously important to simplify the operations in every procedure.

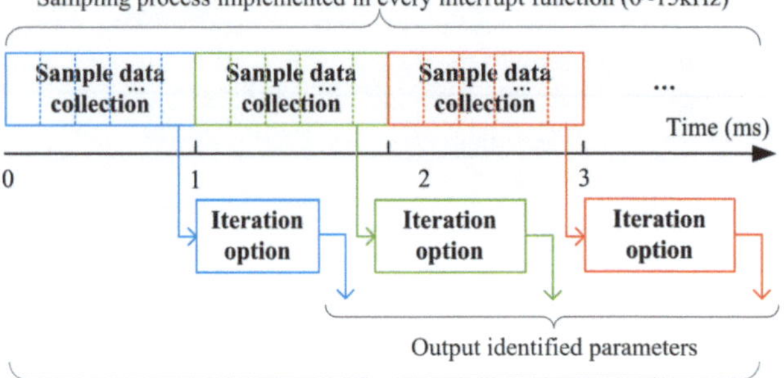



**Figure 5.** Implementation of the PSO-based online inductance identification.

Firstly, the fitness function in the identification of $L_q$ is designed as follows according to (24).

$$
\begin{cases}
F(\hat{L}_q) = \sum_{n=1}^{N} [y(n) - \hat{y}(n)]^2 \\
y(n) = \psi_{ext}(n)^2 - \left[\psi_\alpha(n)^2 + \psi_\beta(n)^2\right] \\
\hat{y}(n) = z_1(n)\hat{L}_q^2 + z_2(n)\hat{L}_q \\
z_1(n) = I_s(n)^2 \\
z_2(n) = -2\left[\psi_\alpha(n)i_\alpha(n) + \psi_\beta(n)i_\beta(n)\right]
\end{cases}
\tag{27}
$$

where $n$ is the number of sample points, $1 \leq n \leq N$ and $N$ is the sum of sample points. And $y(n), z_1(n), z_2(n)$ are the sample data, the calculation of which is simple enough to be finished along with the sampling process. As a result, only two equations in (27) need to be calculated in the identification process.

Secondly, the sampling frequency is generally the same as the switching frequency of the converter, which is usually 6~15 kHz. That is to say, the sum of sample points is 6~15. However, the current, voltage and flux linkage of the PMSM would not vary dramatically in 1 millisecond, so it is probable that there will be similar sample data values. A pretreatment process is added to avoid repetitive operations caused by similar sample points. Specifically, if the difference in the current, voltage and flux linkage are within 1% of the rated value, these sample points will be regarded as the same.

Furthermore, the actual varying law of $L_q$ can offer help with accelerating the convergence velocity. The initialization of traditional PSO is completely random. However, the variation in the inductance is sequential and bounded. It is reasonable to keep the offline measurement result and the last identification result as two of the initial particle locations. And the searching range is set between 20% and 200% of the offline measurement result. Furthermore, as the inductance is negatively correlated with the stator current, the initial velocity of the particles can be set randomly as a positive value if the stator current decreases and vice versa.

All the above measures can effectively improve the efficiency of PSO, and the modified online identification process based on PSO is described in Figure 6.

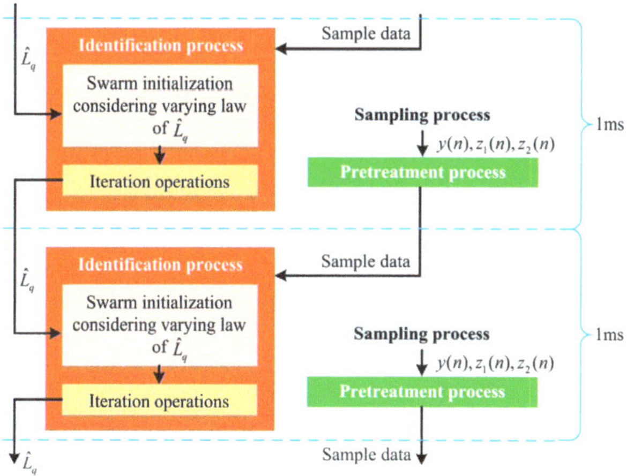


**Figure 6.** Modified online identification process of $L_q$ with efficiency promotion measures.

Even with the modified process, it is difficult to guarantee the identification can be finished within 1 millisecond. And the most effective approach is reducing the number of particles and iterations if possible. Simulations of the online identification of $L_q$ with different number of particles and iterations are conducted to evaluate this issue. The number of particles is set to 10 and 20, while the number of iterations is set to 5, 10, 15 and 20, respectively. The simulation results are given in Table 2. It can be found that with more particles and iterations, more accurate and stable identification results are achieved. However, with the mentioned settings for the searching range and methods, the difference in the identification error among different iterations is not very large. More precisely, the average error of 5 iterations only increases by 2% compared with that of 20 iterations. Indeed, Figure 7 shows the identification process with different number of particles, where each identification result is obtained after five iterations. Though the identification process with 20 particles produces better results, the error with 10 particles is not unacceptable in consideration of the reduced computation burden.

**Table 2.** Simulation results of the proposed PSO-based $L_q$ identification with 10 particles.

| Iterations | Mean Relative Error with 10 Particles (%) | Mean Relative Error with 5 Particles (%) | Relative Standard Deviation with 10 Particles (%) | Relative Standard Deviation with 5 Particles (%) |
|---|---|---|---|---|
| 5 | −0.80 | −0.47 | 3.23 | 3.15 |
| 10 | −0.67 | −0.50 | 3.29 | 3.17 |
| 15 | −0.66 | −0.48 | 3.19 | 3.10 |
| 20 | −0.62 | −0.46 | 3.16 | 3.05 |

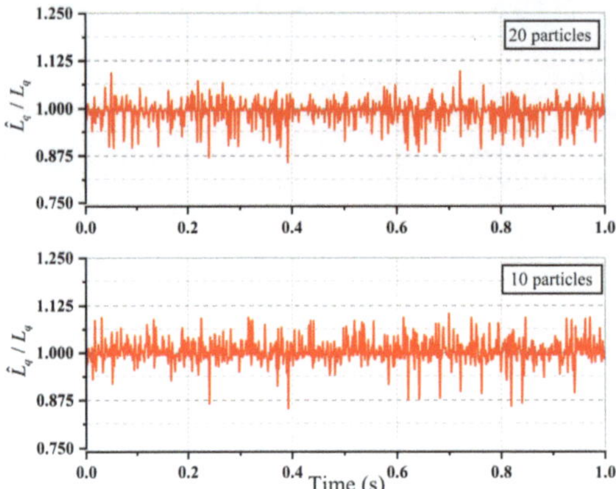

**Figure 7.** Simulation results of the proposed PSO-based $L_q$ identification methods using different numbers of particles after five iterations.

### 4.3. Parametric Iquation Derivation of d-Axis Inductance

The online identification process of $L_q$ is independent of not only the rotor position information but also the d-axis inductance. So, it is reasonable to utilize the identification result of $L_q$ in the identification of $L_d$. Another rotating reference frame is established based on the estimated rotor position, which is shown as the $\gamma$-$\delta$ frame in Figure 8. The mathematical model in the actual rotating reference frame can be transformed into the model in the $\gamma$-$\delta$ frame as follows:

$$\begin{bmatrix} u_\gamma \\ u_\delta \end{bmatrix} = \begin{bmatrix} R_s + pL_d & -\omega_e L_q \\ \omega_e L_q & R_s + pL_d \end{bmatrix} \begin{bmatrix} i_\gamma \\ i_\delta \end{bmatrix} + E_{ext} \begin{bmatrix} -\sin \Delta\theta_e \\ \cos \Delta\theta_e \end{bmatrix} \tag{28}$$

where $\Delta\theta_e$ is the difference between the actual rotor position and the estimated rotor position, and $u_\gamma, u_\delta, i_\gamma, i_\delta$ are the voltage components and current components in the estimated rotating frame. The expression of $u_\delta$ is discretized to the following equation:

$$i_\delta(j+1) = i_\delta(j) + \frac{T_s}{L_d}u_\delta(j) - \frac{T_s}{L_d}R_s i_\delta(j) - \frac{T_s}{L_d}\omega_e(j)L_q i_\gamma(j) - \frac{T_s}{L_d}E_{ext,\delta}(j) \tag{29}$$

where $T_s$ is the sampling period, $j$ is the index of sampling instants and $E_{ext,\delta}$ is the $\delta$-axis component of the extended back EMF, $E_{ext,\delta} = E_{ext} \cos \Delta\theta_e$. And the expression at the last instant is:

$$i_\delta(j) = i_\delta(j-1) + \frac{T_s}{L_d}u_\delta(j-1) - \frac{T_s}{L_d}R_s i_\delta(j-1) - \frac{T_s}{L_d}\omega_e(j-1)L_q i_\gamma(j-1) - \frac{T_s}{L_d}E_{ext,\delta}(j-1) \tag{30}$$

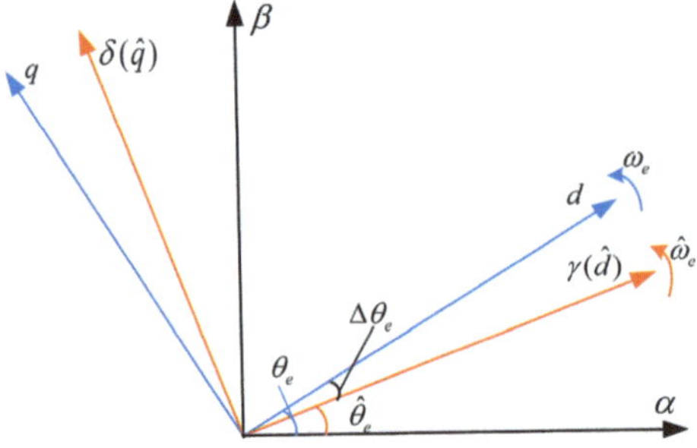

**Figure 8.** Rotating reference frame based on the estimated rotor position.

Compared with the voltage and current components, the variation in the electrical speed, extended back EMF and estimated position error in two contiguous sampling periods can be neglected. Based on the above assumption, the following equation can be obtained by taking the difference between (28) and (29).

$$L_d[\Delta i_\delta(j+1) - \Delta i_\delta(j)] = T_s[\Delta u_\delta(j) - R_s\Delta i_\delta(j) - \omega_e(j)L_q\Delta i_\gamma(j)] \tag{31}$$

with:

$$\begin{cases} \Delta i_\delta(j) = i_\delta(j) - i_\delta(j-1) \\ \Delta i_\gamma(j) = i_\gamma(j) - i_\gamma(j-1) \\ \Delta u_\delta(j) = u_\delta(j) - u_\delta(j-1) \\ \omega_e(j) \approx \omega_e(j-1) \end{cases} \tag{32}$$

Equation (31) is the parametric equation of $L_d$, which is exclusive of rotor position.

### 4.4. Online Identification of d-Axis Inductance Using RLS

Unlike the parametric equation of $L_q$, (31) is a simple linear expression. And the RLS method in Section 3 is adopted to smooth the identification results as follows:

$$\begin{cases} Y(j) = T_s[\Delta u_\delta(j) - R_s\Delta i_\delta(j) - \omega_e(j)L_q\Delta i_\gamma(j)] \\ X(j) = \Delta i_\delta(j+1) - \Delta i_\delta(j) \\ W(j) = L_d(j) \end{cases} \tag{33}$$

It has to be noted that (31) is derived based on the transient model, and the update frequency of $L_d$ is the same as the sampling frequency, which is several times that of the update frequency of $L_q$. As a result, the identification process of $L_d$ shares the same identification result as $L_q$ between each update. Furthermore, the proposed inductance identification method can be combined with various resistance and permanent flux linkage identification methods with the help of the two-timescale structure proposed in [24]. Furthermore, the identification method can be applied along with all kinds of model-based position estimators to construct the sensorless control system. The overall structure of the sensorless control system with the proposed online inductance identification method can be summarized in Figure 9. It should be noted that the resistance and permanent flux linkage used in this manuscript are obtained using offline measurement under different temperatures. The offline measurement results are made into lookup tables and applied in the inductance identification process.

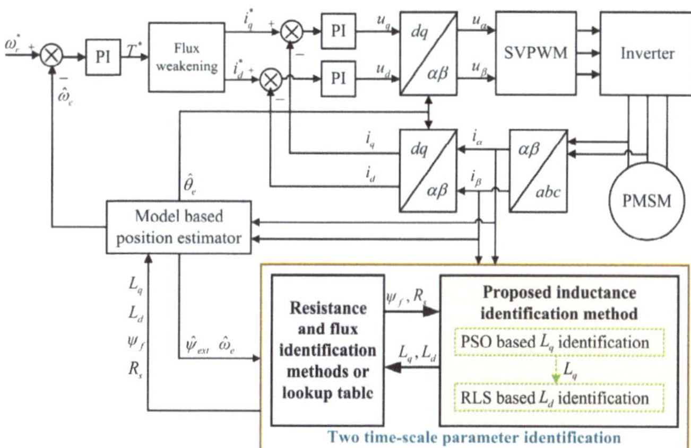


**Figure 9.** The structure of the sensorless control system with online inductance identification (where * denotes the reference values).

Simulation is performed to verify the whole inductance identification method and the results are given in Figure 10. The initial value for the motor inductance is set to 60% of its actual value, which is used to realize sensorless control of the motor in the beginning. The motor operates at its rated conditions and the identification process starts at 0.2 s in Figure 10 to display the effectiveness of identification. The simulation results show that both $L_d$ and $L_q$ can converge to the actual value rapidly. The PSO-based identification of $L_q$ converges faster while the RLS-based identification of $L_d$ has better stability. The estimated speed can follow the actual speed correctly even with inaccurate inductance. However, the estimated position error decreases significantly after adopting the identified inductance, which validates the beneficial effect of the proposed method.

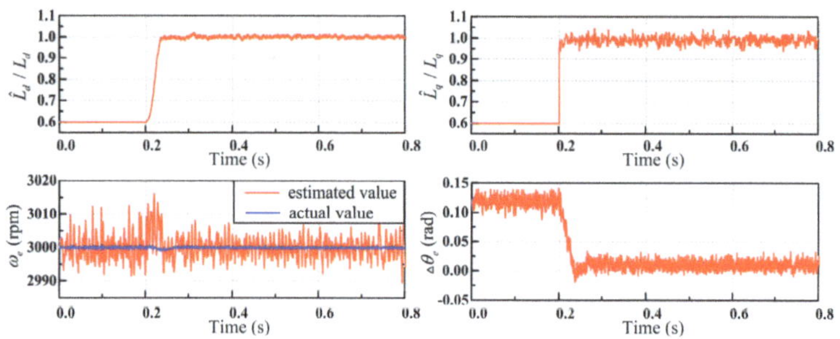

**Figure 10.** Simulation results of the proposed method in sensorless control system of PMSM.

## 5. Experimental Results

The purpose of this section is to verify the properties of the proposed online inductance identification method in a sensorless control system of a PMSM. For the experiments, a homemade three-phase two-level IGBT inverter switching at 5~10 kHz and a DC voltage source of 540 V is applied. The MCU of the inverter is the TC234 by Infineon with a clock speed of 200 MHz. And the proposed inductance identification method is implemented with C language programming based on a product software platform offered by our cooperative enterprise. The tested motor is loaded using a homemade dynamometer. The phase current and voltage are sampled using external LEM LF 1010-S sensors, and the speed and torque of the motor are monitored using a tacho-torquemeter. All the above signals are sent to the Yokogawa WT5000 power analyzer and Yokogawa DLM3024 oscilloscope for further analysis. There is a preassembled resolver in the tested motor, the sampling result of which is sent to the MCU to validate the performance of the proposed method and can used for field-oriented control if necessary. The experimental results from the MCU, oscilloscope and power analyzer are sent to the upper computer through a CAN or Ethernet module. The experimental setup is shown in Figure 11. The parameters of the tested PMSM are the same as those in Table 1.

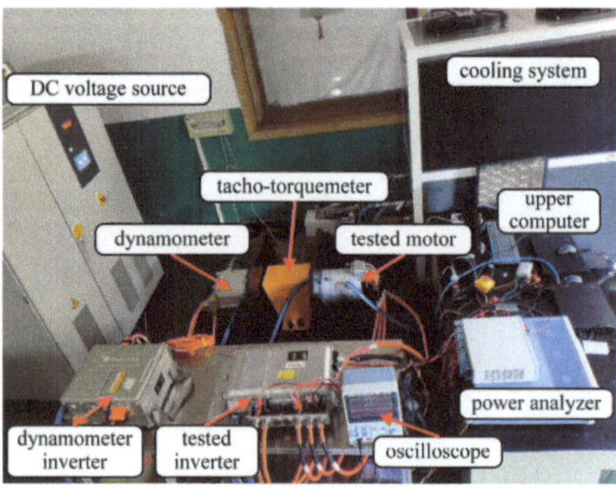


**Figure 11.** The experimental setup.

### 5.1. Execution Time Test

The execution time is always a problem for intelligent algorithms when adopted in online identification methods. As the PSO algorithm is employed in the identification of $L_q$, it is necessary to evaluate the execution time before the bench test. The RLS-based $L_d$ identification process will not suffer from this problem, so only the PSO-based $L_q$ identification process is discussed in this part. The number of sample points is set to 10, and the sample data are prestored in RAM. Then, the PSO-based identification process with different particle numbers and different iterations is executed in the TC234 once per second. The identification process is repeated at least 20 times for each group and the average execution time is calculated and recorded in Table 3.

As mentioned in Section 4, the expected execution frequency of the identification process in this paper is 1 kHz. It is obvious that an identification process with more than 20 particles and 20 iterations can not be accomplished every millisecond. Equally, in order not to affect the original control system, an acceptable execution time for the inductance iden-tification, which is an additional function, should be less than 20% of the total duration. After evaluation of the earlier simulation results and the test results on the execution time, 10 particles and 5 iterations are adopted in the practical applications.

**Table 3.** Execution time of the proposed PSO-based identification in TC234 by Infineon.

| Particles \ Iterations | 5 | 10 | 15 | 20 |
|---|---|---|---|---|
| 10 | 121 µs | 272 µs | 392 µs | 517 µs |
| 15 | 189 µs | 403 µs | 586 µs | 770 µs |
| 20 | 290 µs | 536 µs | 778 µs | 1023 µs |

### 5.2. Experiments under Rated Conditions

The proposed online inductance identification method is employed in the sensorless control system under rated conditions of the motor to verify its effectiveness. The identification results for inductance are adopted in the estimation of the rotor position and speed. The initial value of inductance is set as the nominal value in Table 1. Figure 12 shows the experimental results under a steady state. The identified inductance can settle around the actual value with small jitter, which is obtained using offline parameter measurements with the same currents. The maximum identification error of $L_d$ is 0.025 mH and the maximum identification error of $L_q$ is 0.045 mH, both of which are less than 10% of the actual value. With the identified inductance, the average estimation error of the rotor position is only 0.0334 rad and would not exceed 0.06 rad under a steady state. As a result, the output torque of the motor deviates from the theoretical value by about 2 N·m, which is completely acceptable in a sensorless control system.

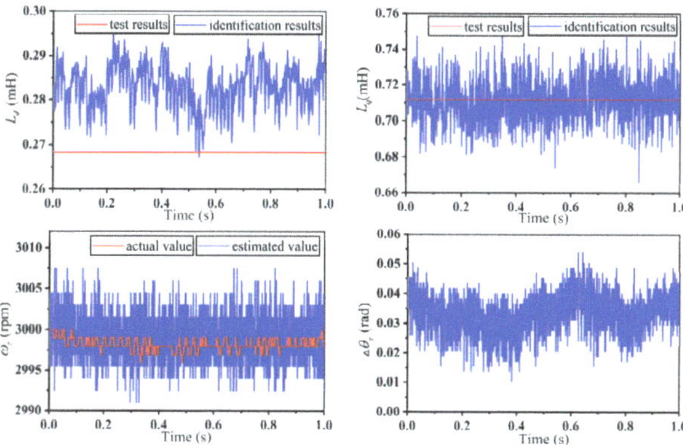


**Figure 12.** Experimental results of the proposed inductance identification method under rated conditions of the motor.

### 5.3. Experiments with Different Initial Values

The inductance identification process is executed with different initial values to verify its convergence situation and speed. The results are given in Figure 13. The identification process starts at 0.1 s. The initial value of the inductance is set to 20%, 50%, 100%, 150% and 200% of the actual value. The identification of $L_q$ would not be affected by the value of $L_d$, and in the experiments on $L_d$ identification, the value of $L_q$ is set to its actual value. Considering that large errors in the inductance will result in the failure of the position estimation, the actual rotor position and speed from the preassembled resolver will be used in the field-oriented control of the motor in this part.

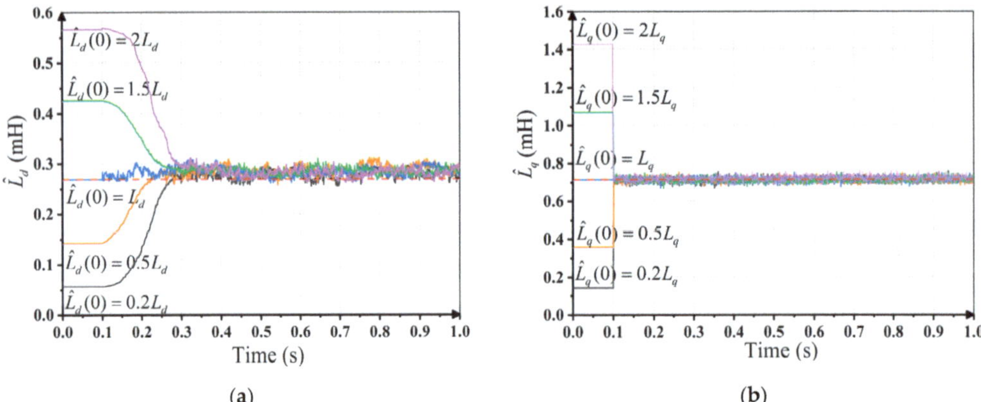


**Figure 13.** Identification results with different initial values under rated conditions of the motor: (**a**) identification results of d-axis inductance; (**b**) identification results of q-axis inductance.

It can be seen from Figure 13 that all of the identification results can converge to the same value. The PSO-based identification of $L_q$ can converge to the actual value in the first few calculations, while the convergence time of the RLS-based identification of $L_d$ is about 0.15~0.2 s with different initial values, which coincides with the simulations. The maximum identification error is less than 10% of the actual value after convergence, showing a similar performance to the experiments in the previous experiments.

### 5.4. Experiments with Different Stator Currents

The inductance of the PMSM can change with the stator current due to self-saturation and cross-saturation effects. The experiments in Section 5.2 are repeated with different stator currents under steady states. As the tested motor runs in the torque control model, changes in current are realized by adjusting the output torque of the motor. The output torque is set between 20 N·m and 220 N·m at an interval of 20 N·m. The relationship between the output torque and stator current is given in Figure 14a. And the average value of the inductance identification results is collected in Figure 14b with the offline identification results as a reference. The offline identification results with the same current are obtained using the standstill frequency–response test, which are described as "the test results" for short in the following contents. It can be found that the identification results are quite close to the test results and the maximum relative identification error is still less than 10%. The identification results present similar changing regulation. The d-axis inductance increases with the output torque while the q-axis decrease gradually as the torque becomes larger. It can be inferred that the saturation effect along the q-axis is enhanced with the output torque. Though the weak magnetic current along the d-axis can reduce the corresponding magnetic saturation effect, the d-axis inductance becomes smaller due to the cross-saturation phenomenon.

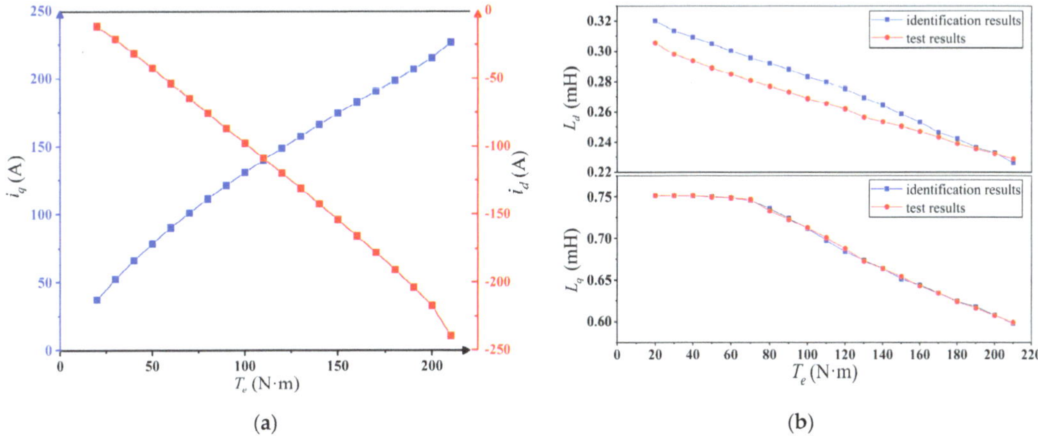

**Figure 14.** Inductance identification results with different output torque: (**a**) corresponding relation between the output torque and the stator current; (**b**) inductance identification results.

In summary, the proposed online inductance identification method is independent of the rotor position information and can be completed within 200 μs in the TC234. The experimental results indicate that the online identification method can converge to the actual value rapidly under a steady state with a maximum relative identification error of 10%. The robustness of the proposed method is verified using different initial values and different stator currents. With the help of the identification results, accurate estimation of the rotor position and speed can be obtained, and the stability and reliability of the sensorless control system can be guaranteed.

## 6. Conclusions

A novel online inductance identification method for a PMSM is proposed in this paper, which is independent of the rotor position information. In the proposed identification method, the parametric equations for the inductance without position information are deduced, and advanced PSO and RLS are utilized to realize real-time identification of $L_d$ and $L_q$, respectively. With the proposed method, accurate inductance identification results can be obtained and employed in the estimation of the rotor speed and position, which further improves the performance of the sensorless control system of the PMSM. The proposed method is efficient enough to be implemented within 200 μs with the help of reasonable simplification. Moreover, the proposed inductance identification method can be adopted in combination with multiple existing resistance and permanent flux linkage identification methods. All the above conclusions have been verified using simulations and experiments. Equally, the experimental results have demonstrated the robustness of the proposed method under different initial values and different stator currents. The proposed online inductance identification method can be employed in sensorless control systems of electric vehicle accessories like pumps, fans and compressors. And it is quite suitable for the fault detection and fault-tolerant control of traction motors in case of the failure of the position sensors in electric vehicles.

**Author Contributions:** Conceptualization, J.X. and J.Z.; methodology, J.X.; software, J.X.; validation, J.X., X.Z. and Y.X.; formal analysis, J.Z.; writing—original draft preparation, J.X.; writing—review and editing, Y.X.; supervision, J.Z.; funding acquisition, J.Z. All authors have read and agreed to the published version of the manuscript.

**Funding:** This research was funded by the National Key R&D Program of China sponsored by Ministry of Science and Technology of the People's Republic of China (MoST), grant number 2022YFB25031013.

**Data Availability Statement:** The data that support the findings of this study are available on request from the corresponding author. The data are not publicly available due to privacy restrictions.

**Conflicts of Interest:** The authors declare no conflict of interest. Author Xingming Zhuang was employed by the company BIT HuaChuang Electric Vehicle Technology Co., Ltd. The remaining authors declare that the research was conducted in the absence of any commercial or financial relationships that could be construed as a potential conflict of interest.

## References


1. Wei, Y.; Ke, D.; Qi, H.; Wang, F.; Rodríguez, J. Multistep Predictive Current Control for Electrical Drives with Adaptive Horizons. *IEEE Trans. Ind. Electron.* **2024**, *71*, 250–260. [CrossRef]
2. Singh, B.; Kashif, M. PF-MRAC-Based Elimination of Sensors in Solar-Powered PMSM Drive-Based Water Pumping System. *IEEE Trans. Ind. Electron.* **2023**, *70*, 11390–11400. [CrossRef]
3. Bariša, T.; Sumina, D.; Pravica, L.; Čolović, I. Flying start and sensorless control of permanent magnet wind power generator using induced voltage measurement and phase-locked loop. *Electr. Power Syst. Res.* **2017**, *152*, 457–465. [CrossRef]
4. Rafaq, M.S.; Jung, J.W. A Comprehensive Review of State-of-the-Art Parameter Estimation Techniques for Permanent Magnet Synchronous Motors in Wide Speed Range. *IEEE Trans. Ind. Inform.* **2020**, *16*, 4747–4758. [CrossRef]
5. Wang, Z.; Yao, B.; Guo, L.; Jin, X.; Li, X.; Wang, H. Initial Rotor Position Detection for Permanent Magnet Synchronous Motor Based on High-Frequency Voltage Injection without Filter. *World Electr. Veh. J.* **2020**, *11*, 71. [CrossRef]
6. Yoon, Y.; Sul, S.; Morimoto, S.; Ide, K. High-Bandwidth Sensorless Algorithm for AC Machines Based on Square-Wave-Type Voltage Injection. *IEEE Trans. Ind. Appl.* **2011**, *47*, 1361–1370. [CrossRef]
7. Jansson, M.; Harnefors, L.; Wallmark, O.; Leksell, M. Synchronization at startup and stable rotation reversal of sensorless non-salient PMSM drives. *IEEE Trans. Ind. Electron.* **2006**, *53*, 379–387. [CrossRef]
8. Kim, Y.W.; Sul, S.K. Stability Analysis of Active Front End and Permanent Magnet Synchronous Generator with Back EMF-Based Sensorless Control for DC Marine Vessels. *IEEE Tran. Power Electron.* **2023**, *38*, 5411–5421. [CrossRef]
9. Boldea, I.; Paicu, M.C.; Andreescu, G.D.; Blaabjerg, F. Active Flux DTFC-SVM Sensorless Control of IPMSM. *IEEE Trans. Energy Convers.* **2009**, *24*, 314–322. [CrossRef]
10. Jukić, F.; Pravica, L.; Sumina, D.; Erceg, I. Framework for Sensorless Control and Flying Start of a Permanent Magnet Generator Based on a Sliding Mode Observer. *IEEE Trans. Ind. Electron.* **2024**, *71*, 294–304. [CrossRef]
11. Quang, N.K.; Hieu, N.T.; Ha, Q.P. FPGA-Based Sensorless PMSM Speed Control Using Reduced-Order Extended Kalman Filters. *IEEE Trans. Ind. Electron.* **2014**, *61*, 6574–6582. [CrossRef]
12. Xu, Y.; Yao, M.; Sun, X. Overview of Position-Sensorless Technology for Permanent Magnet Synchronous Motor Systems. *World Electr. Veh. J.* **2023**, *14*, 212. [CrossRef]
13. Boldea, I.; Paicu, M.C.; Andreescu, G.D. Active Flux Concept for Motion-Sensorless Unified AC Drives. *IEEE Trans. Power Electron.* **2008**, *23*, 2612–2618. [CrossRef]
14. Bolognani, S.; Calligaro, S.; Petrella, R. Design Issues and Estimation Errors Analysis of Back-EMF-Based Position and Speed Observer for SPM Synchronous Motors. *IEEE J. Emerg. Sel. Top. Power Electron.* **2014**, *2*, 159–170. [CrossRef]
15. Nahid-Mobarakeh, B.; Meibody-Tabar, F.; Sargos, F.M. Mechanical sensorless control of PMSM with online estimation of stator resistance. *IEEE Trans. Ind. Appl.* **2004**, *40*, 457–471. [CrossRef]
16. Ahn, H.; Park, H.; Kim, C.; Lee, H. A review of state-of-the-art techniques for PMSM parameter identification. *J. Electr. Eng. Technol.* **2020**, *15*, 1177–1187. [CrossRef]
17. Bolognani, S.; Ortombina, L.; Tinazzi, F.; Zigliotto, M. Model sensitivity of fundamental-frequency-based position estimators for sensorless pm and reluctance synchronous motor drives. *IEEE Trans. Ind. Electron.* **2018**, *65*, 77–85. [CrossRef]
18. Wang, Q.; Zhang, G.; Wang, G.; Li, C.; Xu, D. Offline parameter self-learning method for general-purpose PMSM drives with estimation error compensation. *IEEE Trans. Power Electron.* **2019**, *34*, 11103–11115. [CrossRef]
19. Filho, C.J.V.; Xiao, D.; Vieira, R.P.; Emadi, A. Observers for High-Speed Sensorless PMSM Drives: Design Methods, Tuning Challenges and Future Trends. *IEEE Access* **2021**, *9*, 56397–56415. [CrossRef]
20. Huang, Y.; Zhang, J.; Chen, D.; Qi, J. Model Reference Adaptive Control of Marine Permanent Magnet Propulsion Motor Based on Parameter Identification. *Electronics* **2022**, *11*, 1012. [CrossRef]
21. Wu, C.; Zhao, Y.; Sun, M. Enhancing low-speed sensorless control of PMSM using phase voltage measurements and online multiple parameter identification. *IEEE Trans. Power Electron.* **2020**, *35*, 10700–10710. [CrossRef]
22. Inoue, Y.; Kawaguchi, Y.; Morimoto, S.; Sanada, M. Performance improvement of sensorless IPMSM drives in a low-speed region using online parameter identification. *IEEE Trans. Ind. Appl.* **2011**, *47*, 798–804. [CrossRef]
23. Ichikawa, S.; Tomita, M.; Doki, S.; Okuma, S. Sensorless control of permanent-magnet synchronous motors using online parameter identification based on system identification theory. *IEEE Trans. Ind. Electron.* **2006**, *53*, 363–372. [CrossRef]
24. Rafaq, M.S.; Mwasilu, F.; Kim, J.; Choi, H.H.; Jung, J.-W. Online parameter identification for model-based sensorless control of interior permanent magnet synchronous machine. *IEEE Trans. Power Electron.* **2017**, *32*, 4631–4643. [CrossRef]
25. Colombo, L.; Corradini, M.L.; Cristofaro, A.; Ippoliti, G.; Orlando, G. An embedded strategy for online identification of PMSM parameters and sensorless control. *IEEE Trans. Control Syst. Technol.* **2019**, *27*, 2444–2452. [CrossRef]

26. Ye, S.; Yao, X. A modified flux sliding-mode observer for the sensorless control of PMSMs with online stator resistance and inductance estimation. *IEEE Trans. Power Electron.* **2020**, *35*, 8652–8662. [CrossRef]
27. Gwangmin, P.; Kim, G.; Gu, B. Sensorless PMSM Drive Inductance Estimation Based on a Data-Driven Approach. *Electronics* **2021**, *10*, 791–801.
28. Sun, X.; Zhang, Y.; Tian, X.; Cao, J.; Zhu, J. Speed sensorless control for IPMSMs using a modified MRAS with gray wolf optimization algorithm. *IEEE Trans. Transp. Electrif.* **2022**, *8*, 1326–1337. [CrossRef]
29. Foo, G.; Rahman, M.F. Sensorless Sliding-Mode MTPA Control of an IPM Synchronous Motor Drive Using a Sliding-Mode Observer and HF Signal Injection. *IEEE Trans. Ind. Electron.* **2010**, *57*, 1270–1278. [CrossRef]
30. Hasegawa, M.; Matsui, K. Position sensorless control for interior permanent magnet synchronous motor using adaptive flux observer with inductance identification. *IET Electr. Power Appl.* **2009**, *3*, 209–217. [CrossRef]
31. Chen, Z.; Tomita, M.; Doki, S.; Okuma, S. An extended electromotive force model for sensorless control of interior permanent magnet synchronous motors. *IEEE Trans. Ind. Electron.* **2003**, *50*, 288–295. [CrossRef]
32. Zhang, J.; Peng, F.; Huang, Y.; Yao, Y.; Zhu, Z. Online Inductance Identification Using PWM Current Ripple for Position Sensorless Drive of High-Speed Surface-Mounted Permanent Magnet Synchronous Machines. *IEEE Trans. Ind. Electron.* **2022**, *69*, 12426–12436. [CrossRef]
33. Li, Z.W.; Chen, D.D.; Chen, Y.; Lei, H.D.; Zhu, H.G. PMSM parameter identification based on improved PSO. *J. Phys. Conf. Ser.* **2021**, *1754*, 12235. [CrossRef]
34. Liu, W.; Liu, L.; Chung, I.-Y.; Cartes, D.A. Real-time particle swarm optimization based parameter identification applied to permanent magnet synchronous machine. *Appl. Soft. Comput.* **2011**, *11*, 2556–2564. [CrossRef]

*Article*

# Combined Electromagnetic and Mechanical Design Optimization of Interior Permanent Magnet Rotors for Electric Vehicle Drivetrains

**Guanhua Zhang and Geraint Wyn Jewell ***

Department of Electronic and Electrical Engineering, University of Sheffield, Sheffield S1 3JD, UK; gzhang19@sheffield.ac.uk
* Correspondence: g.jewell@sheffield.ac.uk

**Abstract:** In many high-speed electrical machines, centrifugal forces within the rotor can be first-order constraints on electromagnetic optimization. This can be particularly acute in interior permanent magnet (IPM) machines in which magnets are usually retained entirely by the rotor core with no additional mechanical containment. This study investigates the nature of the trade-off between mechanical and electromagnetic requirements within the context of an eight-pole, 100 kW IPM machine with a base speed of 4000 rpm and an extended speed range up to 12,000 rpm. A series of mechanical and electromagnetic models are used to estimate the level of shaft interference, mechanical stress in critical regions of the rotor and the impact of various features and dimensions within the machine on electromagnetic torque. A systematic exploration of the design space is undertaken for rotor diameters from 120 mm to 180 mm, with optimal designs in terms of torque per unit length established at each diameter while meeting the constraints imposed on mechanical stress. The final preferred design has a rotor of 165 mm and an axial length of 103 mm long with a fractional slot winding in a 30-slot stator. The overall machine has an active mass of 42.3 kg, which corresponds to ~2.4 kW/kg. This paper describes the optimization study in detail and draws on the results to explore the nature of the design trade-offs in such rotors and the impact of core properties.

**Keywords:** permanent magnet machines; rotor mechanical design; interior permanent magnet rotors

**Citation:** Zhang, G.; Jewell, G.W. Combined Electromagnetic and Mechanical Design Optimization of Interior Permanent Magnet Rotors for Electric Vehicle Drivetrains. *World Electr. Veh. J.* **2024**, *15*, 4. https://doi.org/10.3390/wevj15010004

Academic Editor: Joeri Van Mierlo

Received: 20 October 2023
Revised: 13 November 2023
Accepted: 17 November 2023
Published: 21 December 2023



## 1. Introduction

Permanent magnet (PM) machines have become the favored machine technology for the vast majority of electric vehicle (EV) drivetrains due to their combination of superior power density, power factor and efficiency at rated power when compared to competing machine technologies [1]. In permanent magnet machines whose rotors are not subjected to large centrifugal loads, there is a generally accepted maxim that torque density tends to improve with increasing rotor diameter [2]. However, in some applications, the combination of diameter and rotational speed results in large centrifugal loads within the rotor. There is no definitive threshold beyond which machines are regarded as high-speed machines, although peripheral linear speeds beyond 100 m/s or so have been proposed as a useful definition [3]. In most high-speed machines, the mechanical loads may be sufficient to have a marked impact on the mechanical design of the rotor, and, hence, there is often a need to trade off some electromagnetic performance to ensure a robust rotor mechanical design. Moreover, the mechanical load tends to push the optimum aspect ratio to achieve a given torque rating towards longer- and smaller-diameter machines [2].

Interior permanent magnet (IPM) machines in which individual magnet poles are fully enclosed within matching apertures in the rotor core have been adopted in many EV drivetrains and, indeed, are the preferred rotor topology in most mainstream EV applications other than very-high-speed machines in niche vehicle applications. This is due to a combination of their ability to be operated readily in so-called 'field-weakening' mode, the scope to produce saliency torque to supplement the PM excitation torque and the scope for an IPM rotor core to provide the sole mechanical containment and support for the rotor PMs, thus simplifying rotor manufacture. If the centrifugal loading produced through a combination of the self-loading of the core and dead-weight loading from the PMs can be accommodated within the stress limits of the rotor cores, then there is no requirement to employ a separate sleeve or can on the rotor [4]. Published design studies have shown that V-shaped IPM topologies tend to have superior performance compared to other IPM topologies, require less magnet volume and can withstand higher mechanical stresses [5–7].

However, to exploit the self-containing capability of IPM rotor cores, it may prove necessary to thicken up critical regions of the rotor core to take the centrifugal load at acceptable levels of localized stress. The narrow regions adjacent to the airgap that close off the apertures that contain the PMs tend to be the regions of the highest mechanical stress and that tend to require thickening up to accommodate high-speed operation. However, these are also the regions that tend to promote the leakage of the PM flux within the rotor. However, thickening up these regions to accommodate centrifugal loading tends to have an adverse effect on the magnitude of the airgap flux density and, hence, the torque capability. Hence, for a given combination of torque and speed, there is likely to be an optimum rotor size and aspect ratio that yields the maximum torque density. This study seeks to investigate this design trade-off in IPM machines through the electromagnetic and mechanical optimization of the rotor of an IPM machine. Although it is focused on a particular performance specification, many of the trade-offs involved have a more generic utility.

## 2. Design Study Baseline Machine Dimensions and Performance

The performance specifications of the IPM machine considered in this study are representative of a traction machine for a medium-sized, mass-market passenger vehicle, and they are summarized in Table 1.

**Table 1.** Machine performance specifications.

| Feature | Value |
| --- | --- |
| Rated power at base speed | 100 kW |
| Base speed | 4000 rpm |
| Maximum speed | 12,000 rpm |
| Rated torque at base speed | 239 Nm |

The rotor geometry adopted is based on a single-layer V-shaped arrangement of magnets, as shown in Figure 1. The individual geometry of the magnet pieces is profiled at their outer edges, as shown in Figure 2, in an attempt to reduce mechanical stress concentrations in the surrounding rotor core. It is recognized that this intricate shaping of the magnets has cost implications, but as shown later in this paper, this profiling yields dividends in terms of allowing the rotor core to be shaped to reduce leakage flux within the rotor, in turn enhancing torque. The stator and rotor core are NO20 silicon–iron, and the rotor permanent magnets are 35MGOe NdFeB grade with a room temperature remanence of 1.22 T.

The baseline machine design has an 8-pole rotor and a 30-slot stator core, which is equipped with a fractional slot winding with the coil arrangement shown in Figure 1. The machine cross-section in Figure 1 is scaled for various rotor diameters in a systematic manner, which maintains the proportions shown in Figure 1. Table 2 shows the leading dimensions of the machine normalized by the rotor diameter. The normalized dimensions shown correspond to a so-called split ratio [8] of 0.64, i.e., the ratio of the rotor outer diameter to the stator outer diameter.

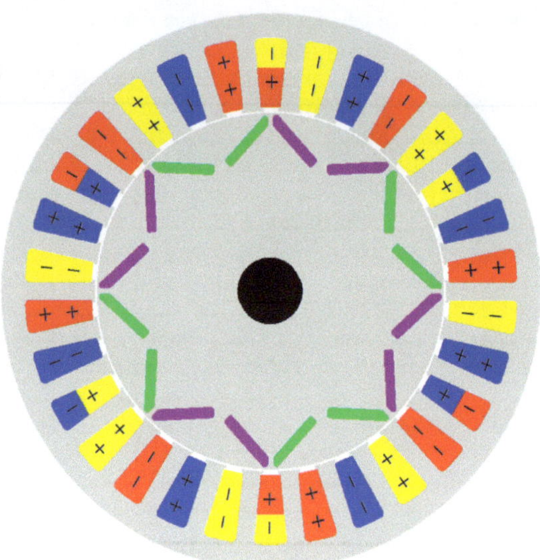

**Figure 1.** Cross-section through the baseline IPM machine.

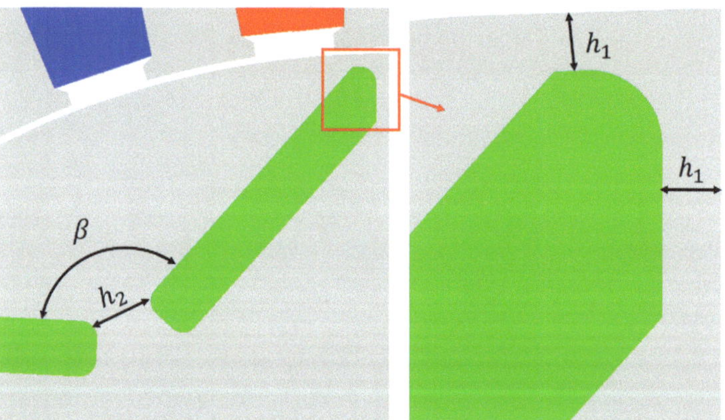


**Figure 2.** Key dimensions around the individual magnet poles.

**Table 2.** Main machine dimensions normalized by rotor outer diameter.

| Feature | Value |
| --- | --- |
| Rotor outer diameter | 1.000 |
| Stator outer diameter | 1.556 |
| Magnet thickness (in direction of magnetization) | 0.0296 |
| Stator tooth body width | 0.0504 |
| Shaft diameter | 0.185 |

In all cases, the stator tooth spans an angle of 7° at the airgap, and the angle between the magnets that make up one pole (defined as $\beta$ in Figure 2) is 130°. The three radii on the magnet corners shown in Figure 2 are set to 1 mm for the 135 mm diameter machine and then scaled in accordance with the method for other machine dimensions, with a maximum radius of 1.3 mm in the case of the 180 mm diameter rotor. The only modification to a pure scaling of the machine cross-section with the rotor diameter is a small adjustment to the stator bore diameter to maintain the same 0.5 mm mechanical airgap in all designs.

## 3. Material Properties and Mechanical Properties

### 3.1. Magnetic Material Properties

The microstructure and processing of most electrical steels are optimized almost entirely to achieve the highest magnetic functional properties. As a result, the mechanical properties of electrical steels tend to be modest in comparison to those of steels or other alloys optimized solely for mechanical behavior. There are often multiple grades within the product range of a given manufacturer, and they offer some trade-off between mechanical and magnetic properties [9]. Although, in principle, this offers the opportunity to employ different grades of electrical steels in the rotor and stator cores, the utilization of electrical steels is generally poor when a single-piece stator alone is taken out of the starting strip of the material, with the region within the bore of the stator core being discarded as scrap. Improved utilization of the starting strip can be realized by taking out the rotor core from within the bore of the stator core. This may result in the use of a grade of electrical steel for the rotor core that has some trade-off in mechanical and magnetic properties to balance the different demands of the stator and rotor cores.

This study is based on NO20-1200H core material, which is a specific grade of 0.2 mm thick, non-oriented silicon–iron. This is not a grade developed specifically for its mechanical properties, but by virtue of being 0.2 mm thick, it is often adopted as a mainstream option for the stators of higher-speed machines. The manufacturer quoted that the minimum 0.2% yield strength is 400 MPa in the rolling direction, which increases by ~2% in the transverse direction. One important consideration is the margin adopted between the design stress limit and the material yield stress. Setting this margin requires consideration of factors such as fatigue, temperature cycling and aging effects. Typically, a design stress limit of 50–60% of the manufacture-quoted yield strength would be adopted, although with extensive in-service experience and operational data, this percentage might be increased. Adopting a design stress limit that is 60% of the 0.2% yield strength of NO20 gives rise to a design stress of 240 MPa for this optimization study. The key mechanical properties adopted for the NO20 rotor core and the NdFeB rotor magnets are summarized in Table 3.

**Table 3.** Rotor material mechanical properties.

| Feature | NO20-1200H | NdFeB |
|---|---|---|
| Strength | 400 MPa (0.2% Yield) | 80 MPa (UTS) |
| Young's modulus | 205 GPa | 160 GPa |
| Poisson's ratio | 0.3 | 0.3 |
| Density | 7650 kg/m$^3$ | 7500 kg/m$^3$ |

### 3.2. Influence of Interference Fitting of Shaft

Interference fitting is widely used in powertrain components as a straightforward, reliable and low-cost means of securing shafts into hubs and cores [10]. The degree of interference employed must be carefully specified to ensure that the radial pressure exerted on the shaft by the rotor core is sufficient to transmit the electromagnetic torque. This condition must be met over the full speed and temperature range encountered in service and is usually set with a significant spare margin. However, setting the level of interference must also take account of the fact that the interference fit at the bore of the rotor core increase the stress in the remainder of the rotor core. Hence, it is important not to over-specify the interference.

A useful estimate of the minimum interference required between the shaft and the rotor core can be established from a simplified model in which the rotor is represented as a solid annulus of uniform density. For a rotating solid annulus representation of the rotor core with outer radius $r_{co}$ and inner radius $r_{ci}$, the radial growth at the inner radius for a rotor of density $\rho$, Young's modulus $E$ and Poisson's ratio $v$ when rotating at an angular velocity of $\omega$ is given by Equation (1) from [11]. For the smallest and largest NO20 rotor cores considered in this study, viz., 120 mm and 180 mm outer diameters, the predicted radial growths of a bore that accommodates a 25 mm shaft are 2.2 μm and 4.9 μm, respectively, at a rated speed of 12,000 rpm.

$$\Delta r_{ci} = \frac{\rho \omega^2 r_{ci}}{4E}\left((3+v)r_{co}{}^2 + (1-v)r_{ci}{}^2\right)$$

(1)

Similarly for the shaft, the change in its outer radius when rotating at an angular velocity of $\omega$ is given by

$$\Delta r_{so} = \frac{\rho \omega^2 r_{so}}{4E}\left((1-v)r_{so}{}^2\right)$$

(2)

For a 25 mm diameter steel shaft (with a modulus of 200 GPa and a density of 7800 kg/m$^3$), the resulting radial growth at 12,000 rpm is a mere 0.02 μm. The difference in radial growth between the shaft and the bore of the rotor core in effect reduces the net interference at the maximum speed from the interference set at standstill. The interference at standstill must be sufficient to accommodate the differential radial growth of the rotor core while still maintaining sufficient inward radial pressure on the shaft to transmit the rated torque with an appropriate safety margin. A well-established method for calculating the contact pressure and resulting hoop stresses between interfering concentric cylinders is given in [10]. For the case of a solid shaft, the interference, $\delta_i$, required to achieve a contact pressure between the shaft and the rotor core of $P_i$ is given from [10] by

$$\delta_i = P_i r_{ci}\left(\frac{r_{ci}^2(-1+v_c)}{E_c\left(-r_{co}^2+r_{ci}^2\right)} + \frac{r_{co}^2(1+v_c)}{E_c\left(r_{co}^2-r_{ci}^2\right)} - \frac{r_{ci}^2(-1+v_s)}{E_s r_{ci}^2}\right)$$

(3)

Taking the case of a 180 mm outer diameter rotor and assuming, for the time being, an axial length of 100 mm (an estimate of which can be obtained on an a priori basis for simple torque-sizing equations), the radial pressure required to transmit the rated torque of 239 Nm assuming a worst-case coefficient of friction of 0.2 between the shaft and the core is 12.2 MPa. This, in turn, requires a minimum net radial interference of 1.5 μm at the maximum speed, which, in turn, requires a standstill interference of 6.4 μm to allow for the 4.9 μm radial growth at 12,000 rpm. Allowing a safety factor of 2 on the interference yields a diametrical interference of 25 μm. This, in turn, generates a maximum radial pressure of 99.7 MPa at standstill. The resulting hoop stress in the rotor core at the inner bore is hence given by

$$\sigma_{\theta,core} = \frac{P_i\left(r_{co}^2 + r_{ci}^2\right)}{r_{co}^2 - r_{ci}^2}$$

(4)

This results in a maximum hoop stress within the core at standstill of 103.6 MPa, which is well within the material design limit. Although cautious in terms of the safety margin on the interference required to transmit the torque, a diametrical interference of 25 μm between the core and the shaft is fixed for the remainder of this investigation.

*3.3. Modeling*

Structural finite element calculations were performed using a linear elastic model (i.e., fixed modulus) within the ANSYS environment. The contact faces between the individual magnet pieces and the rotor core were represented as frictional contacts with a coefficient of friction of 0.2. A very light interference fit between the magnet and rotor core of 2 μm was employed, and the interference fit between the rotor core and the shaft was set to 25 μm for all rotor diameter combinations.

A series of rotor outer diameters between 120 mm and 180 mm (in 15 mm increments) were considered. For each rotor diameter, a total of 154 designs were modeled comprising all-inclusive combinations of designs with $h_1$ values from 0.4 mm to 2.5 mm in 0.1 mm steps and $h_2$ values from 0.8 mm to 2 mm in 0.2 mm steps. An example of the resulting von Mises stress distribution in the vicinity of the rotor magnets is shown in Figure 3; this case is a 135 mm diameter rotor at a maximum speed of 12,000 rpm with $h_1$ = 2 mm and

$h_2 = 1$ mm. Although much of the rotor operates at stress levels below 50 MPa, as would be expected, there are regions of stress concentration at the outer tips of the magnets and in the gap between the innermost regions of the magnets. In this case, the peak localized stress is 230 MPa, which is just within the design stress limit of 240 MPa set for the NO20 rotor material.

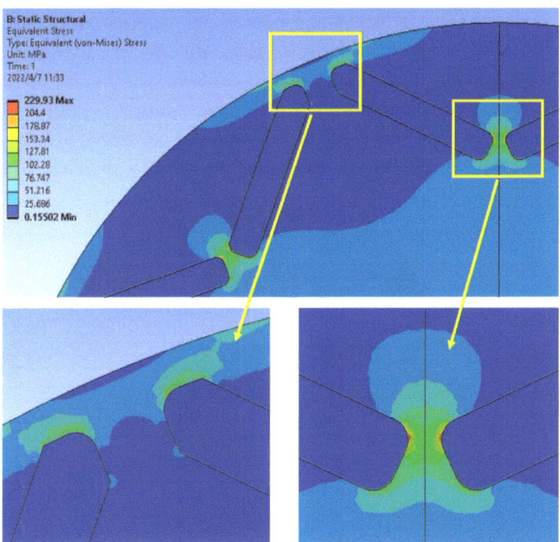



**Figure 3.** IPM von Mises stress distribution for a 135 mm diameter rotor at 12,000 rpm ($h_1 = 2$ mm; $h_2 = 1$ mm).

Taking again the example of the 135 mm diameter rotor, the variation in the maximum predicted localized stress at 12,000 rpm as a function of $h_1$ and $h_2$ for the 154 designs considered is shown in Figure 4. In this case, 40 combinations of $h_1$ and $h_2$ result in localized stress levels greater than the 240 MPa design stress limit set for this study.

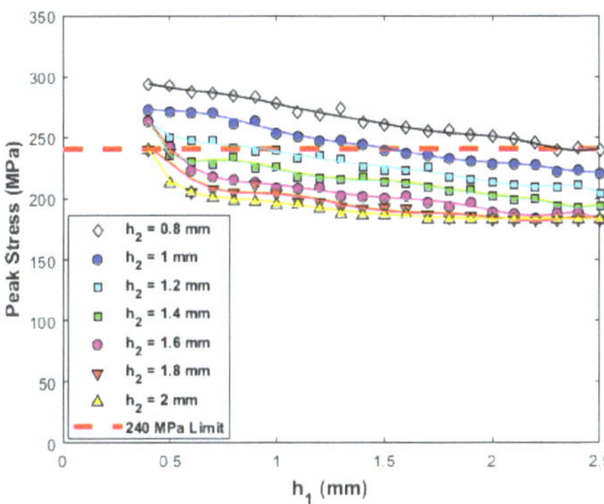



**Figure 4.** Variation in peak localized stress in a series of 135 mm outer diameter rotor designs at 12,000 rpm (core-to-shaft diametrical interference is 25 μm in each case).

Having established the combinations of $h_1$ and $h_2$ that yield viable designs from a mechanical stress threshold perspective, the electromagnetic torque is calculated using a two-dimensional, magneto-static, non-linear finite element analysis for the specific case of a stator rms current density of 10 A/mm$^2$ (at an assumed slot fill factor of 0.45) with a current advance angle of 45° (electrical).

The resulting average torque per unit length values calculated for the 154 design combinations with a 135 mm diameter rotor are shown in Figure 5. Although 40 of these combinations are not viable mechanically, the torques that they produce are included to demonstrate the torque penalty that results from a design stress limit of 240 MPa. For this 135 mm diameter rotor, the highest predicted torque per unit length is 1498 Nm/m for the smallest combination of $h_1$ (0.4 mm) and $h_2$ (0.8 mm), although this results in a peak localized stress of 290 MPa. The highest predicted torque per unit length for a design that also results in a von Mises stress that falls within the design limit of 240 MPa is 1468 Nm/m, which is achieved for an $h_1$ value of 0.5 mm and an $h_2$ value of 1.4 mm. Scaling this torque per unit length to meet the torque specification set out in Table 1 yields an axial length of 162 mm for this rotor diameter of 135 mm, in turn yielding a predicted mass for this design, including an estimate of end-winding mass of 50.6 kg, which corresponds to ~2 kW/kg.

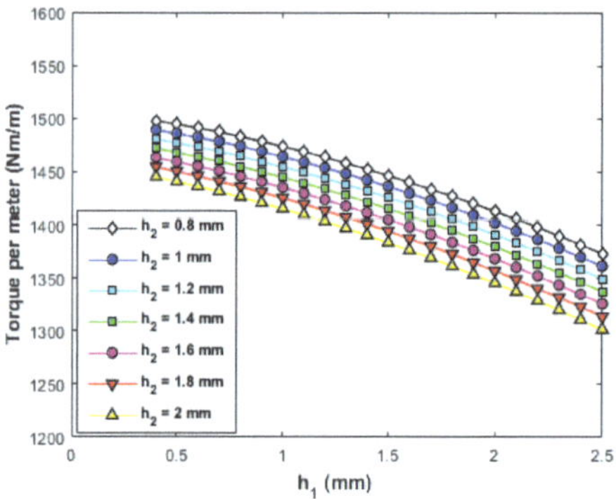


**Figure 5.** Variation in electromagnetic torque in a 135 mm diameter rotor for various combinations of $h_1$ and $h_2$.

Repeating this process to establish the optimal designs for all the rotor outer diameters between 120 mm and 180 mm yields the series of designs summarized in Table 4, which exhibit the maximum torque per unit length within a peak design stress limit of 240 MPa at 12,000 rpm. The axial lengths of the stator and rotor cores of these designs are all scaled to produce a rated torque of 239 Nm at a stator rms current density of 10 A/mm$^2$ (at an assumed slot fill factor of 0.45).

**Table 4.** Optimized designs for different rotor diameters.

|  | D120 | D135 | D150 | D165 | D180 |
|---|---|---|---|---|---|
| $h_1$ (mm) | 0.2 | 0.5 | 0.7 | 1.2 | 3.2 |
| $h_2$ (mm) | 0.7 | 1.4 | 2.9 | 4.6 | 6.9 |
| Core axial length (mm) | 224 | 162 | 125 | 103 | 99 |
| Rotor volume (dm$^3$) | 2.87 | 2.32 | 2.21 | 2.20 | 2.52 |
| Machine mass (kg) | 60.5 | 50.6 | 44.9 | 42.3 | 46.3 |
| Torque density (Nm/kg) | 3.93 | 4.70 | 5.30 | 5.63 | 5.14 |
| Power density (kW/kg) at 4000 rpm | 1.65 | 1.97 | 2.22 | 2.36 | 2.15 |

As would be expected, increasing the rotor diameter beyond 120 mm requires larger values of $h_1$ and $h_2$ to maintain the localized stress within the rotor below the 240 MPa limit. Thickening up the rotor core beyond the end of the magnets tends to promote increased leakage flux within the rotor. It is apparent from the results presented in Table 4 that there is an optimum rotor diameter of 165 mm, albeit that this is specific to this combination of rotational speed, rotor core maximum stress limit, machine split ratio and current density constraints. Without mechanical stress considerations, the torque density tends to continually increase with increasing rotor diameter because of the nature of scaling with the diameter of the electric loading (Ampere turns per unit of airgap periphery). There is a further tendency for the predicted torque density to increase with diameter due to the limitations of two-dimensional finite element modeling. A two-dimensional finite element electromagnetic analysis does not account for the influence of end effects in short-axial-length machines and, hence, tends to overestimate the torque produced by machines with short axial lengths relative to diameters. The presence of an optimum in Table 4 is a consequence of the electromagnetic penalty, which is increasingly incurred with the need to thicken up the regions of the rotor core adjacent to the airgap to ensure that the entire rotor core remains with the specified design mechanical stress.

Figure 6 shows a predicted flux density distribution in the rotor and stator cores of this 165 mm diameter design at the rated torque, while Figure 7 shows a close-up of the von Mises stress distribution in the region around the magnet poles at 12,000 rpm. As is apparent, there is significant magnetic saturation both in regions adjacent to the airgap near the magnet tips and in the regions between the pair of magnets that make up one pole, with peak flux densities of up to ~3.4 T. This saturation plays an important role in limiting the magnet leakage flux within the rotor core, which would otherwise further diminish the airgap flux density. This illustrates the importance of modeling the behavior beyond saturation accurately, as this can have a significant bearing on torque in this machine, even at modest electric loading. In this regard, the representation of the magnetization characteristics is best suited to a semi-analytical model in which saturation is enforced mathematically rather than extrapolated from a series of discrete data points.

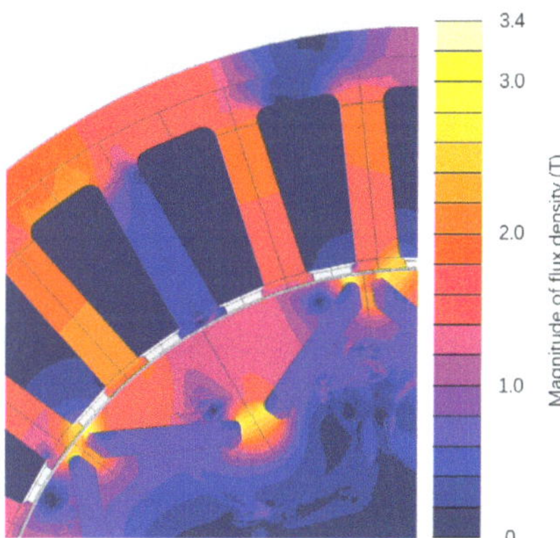

**Figure 6.** Predicted flux density distribution for the 165 mm design in Table 4.

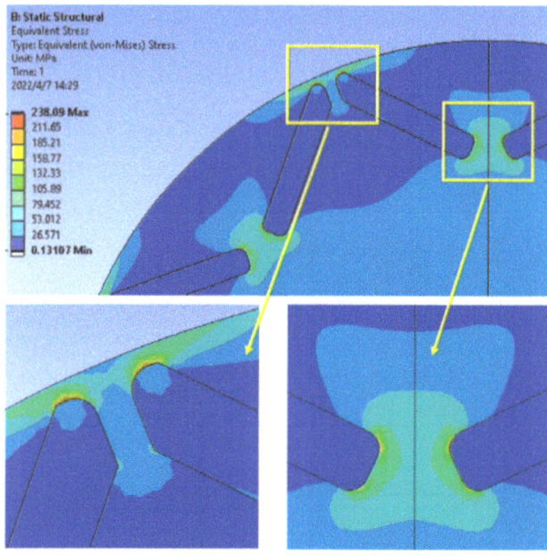

**Figure 7.** Predicted von Mises stress distribution for the 165 mm design in Table 4.

### 4. Influence of Hub Diameter

The rotor designs considered up to this point are all based on scaling the cross-section of the rotor according to a fixed shaft-to-rotor diameter ratio of 0.185. This results in a range of shaft diameters between 22.2 mm and 33.3 mm for the range of rotor outer diameters considered. The remainder of the rotor cross-section is occupied entirely by the rotor core and magnets. From an electromagnetic point of view, much of the inner regions of the rotor need not be magnetic, and, indeed, the nature of the field distribution with an IPM is such that this large region of core within the magnets can promote leakage flux. Hence, employing a non-magnetic and lightweight hub to span the region between the shaft and a slimmed down rotor core will yield dividends in terms of a reduced mass, and it may also yield some electromagnetic benefits. Taking the 165 mm rotor diameter design in Table 4, which offers the highest torque density of all the rotor diameters considered, a series of electromagnetic finite element predictions of average torque are performed with gradually increasing diameter of the non-magnetic hub radius, which, in electromagnetic terms, increases the bore diameter of the rotor core. The non-magnetic hub diameter is limited by the presence of the V-shaped arrangement of the magnets. For the particular arrangement adopted, the inner edges of the magnets are at a diameter of 122 mm for a 165 mm rotor outer diameter.

The resulting variation in the predicted torque with hub diameter is shown in Figure 8; all cases have the same rotor axial length of 103 mm. As is evident and indeed expected, over much of the range considered, the predicted torque remains unchanged from the 239 Nm predicted for the original shaft arrangement. However, as the hub diameter increases beyond 115 mm, the predicted torque starts to increase markedly, with a maximum value of 258 Nm at a rotor core inner diameter of 121 mm. The section of core inside the magnets acts a leakage path for a proportion of the magnet flux, and, hence, as this region is progressively thinned down as the hub diameter increases, this leakage path begins to magnetically saturate. Figure 9 shows the finite element predicted flux density distribution in the rotor at the rated current density for a hub diameter of 120 mm with the 165 mm diameter rotor. The nature of permanent magnet excitation is such that a magnet acts neither as a constant flux source nor a constant source of magneto-motive force. Hence, increasing the reluctance of the leakage path via saturation and a reduced cross-section tends to result in an increasing proportion of the magnet flux crossing the working airgap, hence promoting additional torque production. Hence, from an electromagnetic perspective, there is a meaningful benefit in increasing the hub diameter towards 122 mm. Furthermore, significant mass reductions could be realized if the hub was manufactured from a lightweight metal or composite and/or if features such as spokes or holes were incorporated.

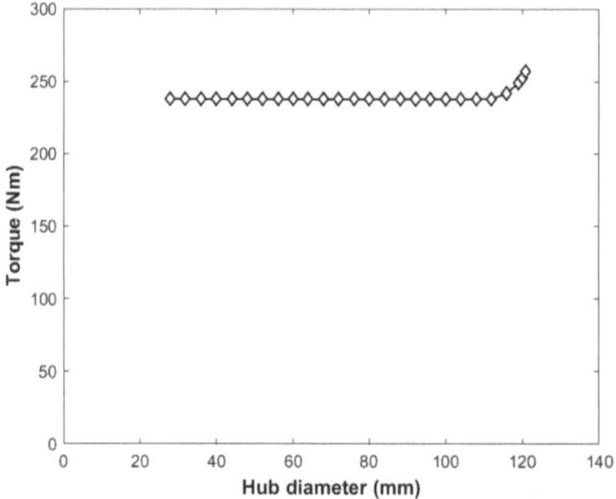

**Figure 8.** Predicted torque for a 165 mm diameter rotor as a function of rotor core inner diameter at a current density of 10 A/mm$^2$ rms (slot fill factor of 0.45).

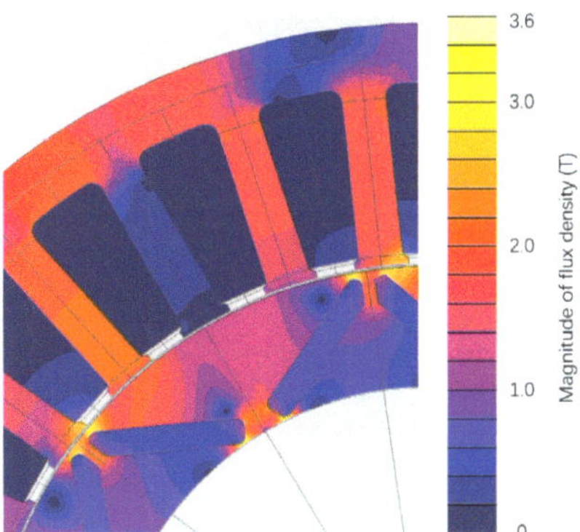

**Figure 9.** Finite element predicted flux density distribution in a 165 mm diameter rotor with a shaft diameter of 120 mm at full load torque of 253 Nm.

However, alongside these electromagnetic and mass considerations, it is necessary to establish the mechanical stress implications of increasing the hub diameter. In the general analytical expressions of the stress in a rotating ring, the hoop stress at the outer edge of the ring of a given outer diameter increases as the bore diameter is increased. Hence, it is to be expected that the hoop stress in the magnet bridge regions of the rotor core will further increase as the bore diameter is increased.

Figure 10 shows the predicted peak stress at 12,000 rpm in the 165 mm diameter rotor, which has the optimum values of $h_1$ and $h_2$ in Table 4 as a function of the hub diameter. As is apparent, there is the expected increase in the peak stress as the hub diameter is increased. Since the original design with a shaft diameter of 30.5 mm is just under the limiting design stress of 240 MPa, it is to be expected that almost all designs with a larger hub diameter will exceed the stress limit. The calculated stress in Figure 10 is shown at four different shaft-to-rotor-core interference levels, from the original 25 μm up to 60 μm. Since the radial growth of the core increases as the hub diameter is increased, there is an upper limit on the hub diameter for a given interference, beyond which the rotor core separates from the hub; e.g., for the original 25 μm, the rotor core lifts of the hub at 12,000 rpm for shaft diameters greater than ~58 mm. Even if the stress could be accommodated for in the 165 mm diameter rotor, it would be necessary to increase the interference to at least 35 μm. This additional interference would further exacerbate the stress levels in the rotor core.

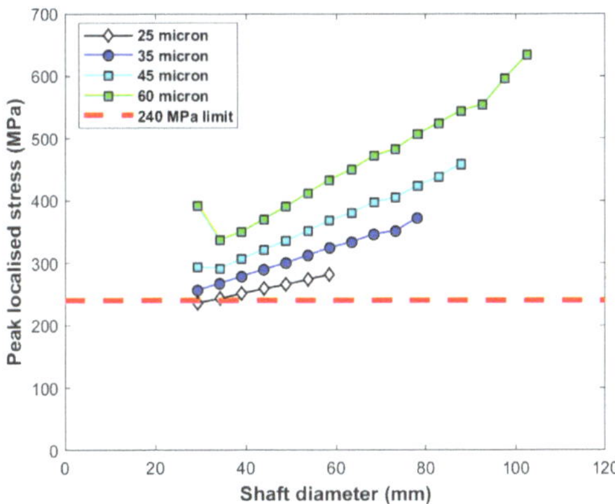


**Figure 10.** Variation in the finite element predicted peak localized stress in the 165 mm diameter rotor core at 12,000 rpm as a function of hub diameter for different diametrical interference fits from 25 μm to 60 μm (240 MPa design stress limit shown as dashed line).

Although almost all the rotor designs in Figure 10 exceed the design stress limit at 240 MPa, it is possible to calculate the limiting speed for each design for an imposed peak stress of 240 MPa. Figure 11 shows the resulting calculated variations in the maximum rotational speed as a function of the hub diameter and interference for a 165 mm diameter rotor.

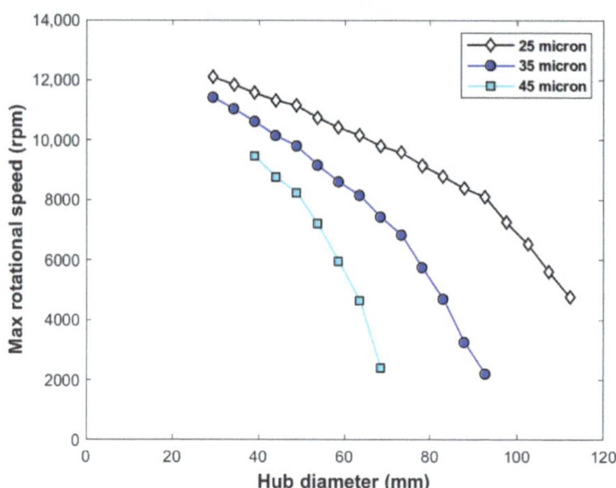


**Figure 11.** Variation in the maximum allowable speed for a 165 mm diameter rotor core to remain within a 240 MPa design stress limit for four levels of interference.

The predicted von Mises stress distribution in the rotor is shown in the close-up in Figure 12 for the particular case of a hub diameter of 78 mm with an initial interference of 35 μm, from which it can be seen that the problem region is again in the bridge regions near the ends of the individual magnet pieces.

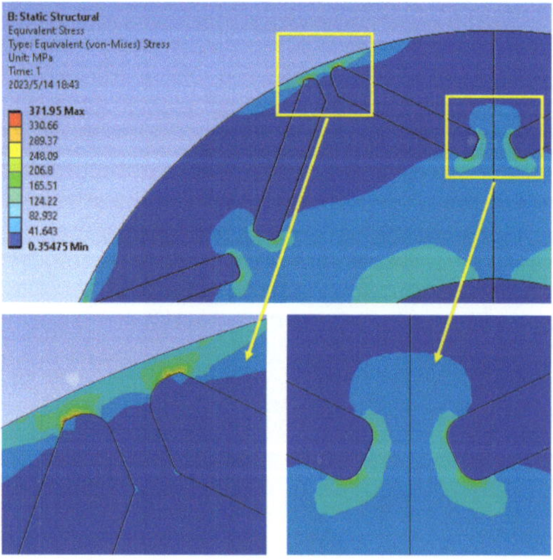


**Figure 12.** Close-up of finite element predicted peak stress distribution in the 165 mm diameter rotor core at 12,000 rpm for a hub diameter of 78 mm.

## 5. Influence of Magnet Shape

The analyses performed up to this point considered magnet poles with profiled detailing at the ends. It was noted earlier in this paper that this profiling was incorporated in an attempt to minimize the localized stress concentrations near the outer edges of the magnets while also minimizing flux leakage. However, this intricate detailing will inevitably increase manufacturing cost and slightly reduce magnet utilization unless the detailing is incorporated though a near-net-shape process. To quantify the performance dividend of incorporating this detailing, a series of further mechanical finite element calculations were performed for a 165 mm diameter rotor with plain rectangular blocks of magnets with the two different rotor core apertures shown in Figure 13. In both cases, the magnet was simply truncated at the ends of the parallel section of the original magnet profile. In the first case shown in Figure 13a, the region previously occupied by the profiled end of the magnet was simply filled by extending the rotor core into this region. In the second variant in Figure 13b, the same region was filled with a non-magnetic filler material, which in practice would be a fiber-loaded epoxy resin or putty, such as HysolEM 300T-KL or Dolphon CV-1108. In both cases, the same 30.5 mm diameter shaft as in the original design in Figure 1 was used, and the optimal combination of $h_1$ and $h_2$ in Table 4 was adopted. The mechanical properties of the non-magnetic filler used in the rotor in Figure 13b will have some influence on the overall stress distribution within the rotor core. To cover the likely range of physical properties exhibited by various loaded resins and putties, a series of simulations were performed assuming that the non-magnetic filler was assigned inclusive combinations of elastic moduli between 2 GPa and 8 GPa and densities between 1200 and 1800 kg/m$^3$. The resulting range of stresses for these different combination of non-magnetic filler properties was rather narrow, with, as expected, the marginally lowest stress of 245 MPa encountered with the lowest density and highest modulus filler and the highest stress of 261 MPa for the highest density and lowest filler.

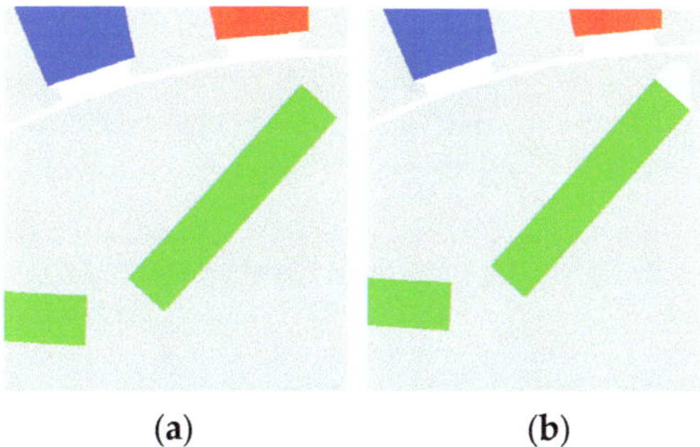

(a)          (b)

**Figure 13.** Alternative arrangement of magnet poles. (**a**) Void from profiled end incorporated into rotor core. (**b**) Void from profiled end filled with non-magnetic filler.

The resulting comparison of the predicted electromagnetic torque between the profiled magnet in Figure 1 and the various rectangular magnets is summarized in Table 5 for the case of a shaft sized according to the scaling factor in Table 2 and a diametric interference fit of 25 μm. As is apparent, the profiled magnet geometry in Figure 1 has the highest predicted torque with a particularly notable advantage over the magnet arrangement in Figure 13a.

**Table 5.** Comparison of predicted electromagnetic torque and maximum localized stress for different magnet profiles.

| | Profiled Magnet in Figure 1 | Magnet in Figure 13a | Magnet in Figure 13b |
|---|---|---|---|
| Torque (Nm) | 239 | 213 | 230 |
| Maximum stress at 12,000 rpm (MPa) | 240 | 229 | 245–261 * |

*—range for combinations of different filler materials modeled.

However, the arrangement in Figure 13a simply replaces the profiled end of the magnet with additional rotor core material, leaving the rectangular magnet at the same location as the rectangular region of the profiled magnet. This inevitably increases flux leakage within the rotor but also serves to reduce the stress in the rotor core due a thickening of the bridging regions. Hence, a more meaningful comparison can be achieved by displacing the rectangular magnet outwards towards the rotor surface, as shown in Figure 14. This tends to reduce the leakage flux within the rotor at the outer edge at the expense of an increased localized mechanical stress. To this end, the rectangular magnet in Figure 13a is progressively moved outwards while maintaining the same value of β (defined previously in Figure 2) until the same peak stress of 240 MPa is achieved.

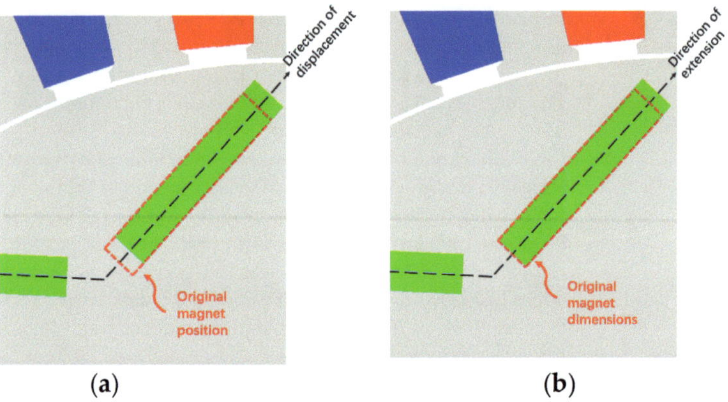


**Figure 14.** Close-up of rectangular magnet blocks. (**a**) Displacing the rectangular magnet in Figure 13a 1.06 mm outwards towards the rotor surface. (**b**) Extending the rectangular magnet in Figure 13a 0.53 mm outwards towards the rotor surface.

The dimensions of the magnet piece itself remain fixed. The resulting rotor geometry is shown in Figure 14, with the magnet moved 1.06 mm along the line shown. This rotor design has an electromagnetic torque of 205 Nm, which is lower than the original rectangular magnet block design in Figure 13a. Although displacing the same magnet block outwards derives some benefits in terms of reducing flux leakage through the bridge regions of the rotor core near the airgap, it does result in an increase in the separation at the inner edges of the two magnets that make up a pole, i.e., the equivalent of dimension $h_2$ in Figure 2. The additional leakage flux through this wider region more than offsets the reduced flux leakage through the outer bridge regions, thus reducing the overall torque.

In an attempt to improve the performance of rotors with plain rectangular magnet blocks, the separation at the inner edges of the two magnets that make up a pole, i.e., the equivalent of dimension $h_2$ in Figure 2, is maintained at its original value of 4.6 mm, and the magnet is simply extended into the region near the outer surface rather than displacing the fixed magnet size. This extension to the rectangular magnet block increases its mass, and, hence, it would be expected that the amount of extension that can be applied before reaching the 240 MPa design stress limit will be less than the displacement of 1.06 mm established previously with the fixed magnet size. For the same stress limit of 240 MPa, the magnet can be extended in the direction shown by 0.53 mm. Figures 15 and 16 show the finite element predicted flux density distribution and von Mises stress distribution at 12,000 rpm for this case of an extended rectangular magnet block, respectively. This extension of the magnet block results in an electromagnetic torque of 218 Nm, which is still 21 Nm short of the 239 Nm achieved by the profiled magnets in Figure 1. Hence, the intricate detailing around the edge of the profiled magnets in Figure 1 results in a ~10% higher torque on a like-for-like basis compared to a plain rectangular block. Bringing the design with the rectangular blocks up to the same torque rating could be achieved with a 10% increase in the stack length and, hence, the mass of the active materials. Further detailed cost modeling would be required to establish whether, in volume production, the intricately profiled magnet could be manufactured within the cost saving margin associated with the 10% reduction in the volume of the active material required.

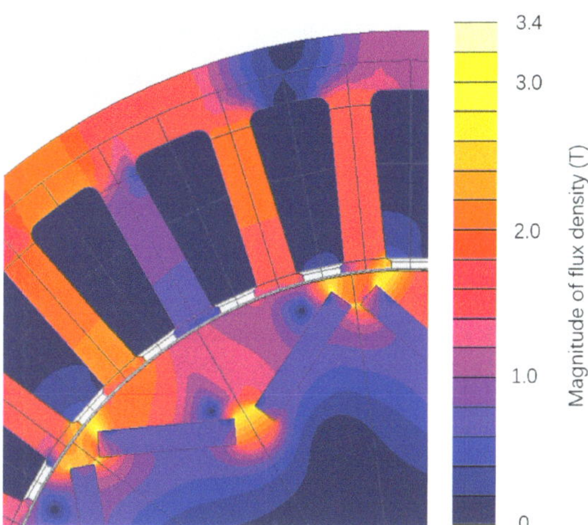


**Figure 15.** Predicted flux density distribution for the 165 mm rotor design with an extended rectangular magnet block.

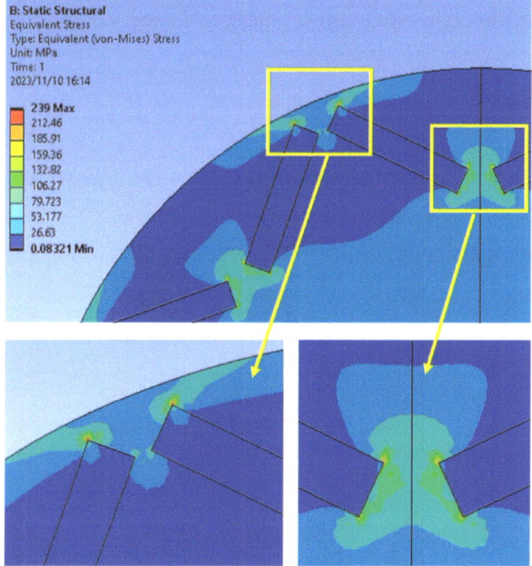


**Figure 16.** Predicted von Mises stress distribution for the 165 mm rotor design with an extended rectangular magnet block.

## 6. Conclusions

This paper described the combined mechanical and electromagnetic optimization of a series of IPM rotors to maximize electromagnetic torque while remaining within a mechanical stress design limit. It showed the critical interplay between stress and the electromagnetic torque capability and that mechanical considerations can result in an optimal rotor diameter, beyond which the compromises in electromagnetic performance that result from accommodating ever-increasing mechanical loads tend to dominate over the natural electromagnetic scaling behavior in which torque density tends to increase with machine size. This study also showed the discernable benefits of profiling the magnets near the outer part of the rotor to reduce stress, in this case resulting in a 10% improvement in torque density compared to a counterpart based on an optimized rectangular block of magnets. This study culminated in an optimized 100 kW IPM machine design with a rotor diameter of 165 mm resulting in an active mass of 42.3 kg, which corresponds to ~2.4 kW/kg.

It is worth noting that the specific findings in terms of torque density and operating speed range are somewhat specific to the specification of a design stress limit of 240 MPa and a current density of 10 A/mm$^2$ rms. The selection of a higher-strength electrical steel for the rotor would improve the torque density due to the ability to accommodate thinner sections within the rotor geometry and, hence, reduce flux leakage within the IPM rotor. However, this might require the use of different electrical steels in the rotor and stator since increased mechanical properties are usually obtained at the expense of reduced magnetic performance, particularly core loss. The 10 A/mm$^2$ rated current density employed throughout this study is representative of many higher-performance machines that employ some level of in-direct cooling of the winding, e.g., liquid-cooled casings or high-performance air cooling. The adoption of higher current densities that are representative of say direct liquid-cooled coils is likely to have a limited influence on the optimal rotor diameter, as it will only affect the armature reaction field contribution to the flux in the bridges within the rotor IPM structure. Although the optimal rotor diameter and resulting torque density would change with the use of higher-strength electrical steels in the rotor and higher current densities in the stator winding, it is likely to be the case that there will be an optimal IPM rotor diameter for a given set of constraints and that establishing this optimum requires consideration of both electromagnetic and mechanical aspects of behavior.

**Author Contributions:** Conceptualization, G.Z. and G.W.J.; methodology, G.W.J.; formal analysis, G.Z.; investigation, G.Z. and G.W.J.; resources, G.W.J.; writing—original draft preparation, G.Z.; writing—review and editing, G.W.J.; supervision, G.W.J. All authors have read and agreed to the published version of the manuscript.

**Funding:** This research received no external funding.

**Data Availability Statement:** Data are contained within the article.

**Conflicts of Interest:** The authors declare no conflict of interest.

## References

1. Bianchi, N.; Bolognani, S.; Luise, F. Potentials and limits of high-speed PM motors. *IEEE Trans. Ind. Appl.* **2004**, *40*, 1570–1578. [CrossRef]
2. Yu, A.; Jewell, G.W. Systematic design study into the influence of rotational speed on the torque density of surface-mounted permanent magnet machines. *J. Eng.* **2019**, *2019*, 4595–4600. [CrossRef]
3. Binder, A.; Schneider, T. High-speed inverter-fed AC drives. In Proceedings of the 2007 International Aegean Conference on Electrical Machines and Power Electronics, Bodrum, Turkey, 10–12 September 2007; pp. 9–16. [CrossRef]
4. Lovelace, E.C.; Jahns, T.M.; Keim, T.A.; Lang, J.H. Mechanical design considerations for conventionally laminated, high-speed, interior PM synchronous machine rotors. *IEEE Trans. Ind. Appl.* **2004**, *40*, 806–812. [CrossRef]
5. Binder, A.; Schneider, T.; Klohr, M. Fixation of buried and surface-mounted magnets in high-speed permanent-magnet synchronous machines. *IEEE Trans. Ind. Appl.* **2006**, *42*, 1031–1037. [CrossRef]
6. Wang, A.; Jia, Y.; Soong, W.L. Comparison of Five Topologies for an Interior Permanent-Magnet Machine for a Hybrid Electric Vehicle. *IEEE Trans. Magn.* **2011**, *47*, 3606–3609. [CrossRef]
7. Han, Z.; Yang, H.; Chen, Y. Investigation of the rotor mechanical stresses of various interior permanent magnet motors. In Proceedings of the 12th International Conference on Electrical Machines and Systems, ICEMS 2009, Tokyo, Japan, 15–18 November 2009. [CrossRef]
8. Ma, J.; Zhu, Z.Q. Optimal split ratio in small high speed PM machines considering both stator and rotor loss limitations. *CES Trans. Electr. Mach. Syst.* **2019**, *3*, 3–11. [CrossRef]

9. Leuning, N.; Schauerte, B.; Hameyer, K. Interrelation of mechanical properties and magneto-mechanical coupling of non-oriented electrical steel. *J. Magn. Magn. Mater.* **2023**, *567*, 170322. [CrossRef]
10. Qiu, J.; Zhou, M. Analytical solution for interference fit for multi-layer thick-walled cylinders and the application in crankshaft bearing design. *Appl. Sci.* **2016**, *6*, 167. [CrossRef]
11. Budynas, R.; Nisbett, J.K. *Shigley's Mechanical Engineering Design*, 10th ed.; McGraw-Hill Education: New York, NY, USA, 2014; ISBN 9813151005.

*World Electric Vehicle Journal*

*Article*

# Speed Stability and Anti-Disturbance Performance Improvement of an Interior Permanent Magnet Synchronous Motor for Electric Vehicles

Zhongxian Chen [1,*], Xianglin Dai [1] and Munawar Faizan [2]

[1]  School of Intelligence Manufacturing, Huanghuai University, Zhumadian 463000, China; daixianglin@huanghuai.edu.cn
[2]  Department of Mechatronic Engineering, Atlantic Technological University, 999014 Galway, Ireland; Faizan.Munawar@research.atu.ie
*  Correspondence: chenzhongxian@huanghuai.edu.cn

**Abstract:** To enhance the speed stability and anti-interference performance of the interior permanent magnet synchronous motor (IPMSM) in electric vehicles, a composite control strategy, incorporating sliding mode control (SMC) and extended state observer (ESO), was implemented to regulate the IPMSM's speed. Firstly, three simulation analysis models of the IPMSM were established based on its electrical parameters. The current-loop regulator was a PI regulator, while the speed-loop regulators consisted of a basic SMC regulator, a linear SMC–ESO regulator, and a nonlinear SMC–ESO regulator. The simulation analysis results demonstrated that all three speed-loop regulators effectively ensured the speed stability of the IPMSM. However, the nonlinear SMC–ESO regulator exhibited superior performance in terms of enhancing the IPMSM's resistance to disturbances. Secondly, a hardware testing platform was constructed to validate the simulation analysis findings. The hardware testing results, when compared to the simulation analysis results, revealed the need for optimization of the PI regulator's control parameters to maintain the speed stability of the IPMSM. Moreover, contrary to the simulation analysis results, the hardware testing results indicated minimal difference in the anti-disturbance performance of the IPMSM between the linear SMC–ESO regulator and the nonlinear SMC–ESO regulator. Finally, the differences between the simulation analysis results and the hardware testing results are thoroughly discussed and analyzed, providing valuable insights for the practical implementation of IPMSM in electric vehicle drive systems.

**Keywords:** speed control; anti-interference performance; interior permanent magnet synchronous motor; extended state observer; electric vehicle

**Citation:** Chen, Z.; Dai, X.; Faizan, M. Speed Stability and Anti-Disturbance Performance Improvement of an Interior Permanent Magnet Synchronous Motor for Electric Vehicles. *World Electr. Veh. J.* **2023**, *14*, 311. https://doi.org/10.3390/wevj14110311

Academic Editor: Ghanim A. Putrus

Received: 23 October 2023
Revised: 7 November 2023
Accepted: 14 November 2023
Published: 16 November 2023

## 1. Introduction

Currently, the permanent magnet synchronous motor (PMSM) for electric vehicles has garnered attention from both scholars and manufacturers [1,2]. The operational stability and safety of electric vehicles rely heavily on the speed regulation (specifically, speed stability) and anti-disturbance performance of the PMSM. Therefore, a comprehensive study into the speed regulation of the PMSM is necessary to enhance the driving stability of electric vehicles. Simultaneously, the anti-disturbance performance of the PMSM plays a pivotal role in ensuring rapid response capability and stability of electric vehicles [3]. Building upon the development of nonlinear control theory, various approaches such as adaptive control, fuzzy neural network control and predictive control have been employed to address the speed regulation and anti-disturbance performance of the PMSM [4–7]. However, the implementation of the nonlinear control theory through computer programs necessitates high-speed computing performance from the hardware controller. Despite these efforts, the speed fluctuation of the PMSM persists due to the inherent nature of nonlinear control theory. Hence, further exploration and research are required to advance the speed stability and anti-disturbance performance of the PMSM.

The Active Disturbances Rejection Control (ADRC) strategy has been widely employed in the enhancement of speed control and anti-disturbance performance of PMSM. By utilizing an ADRC system comprising an Expansion State Observer (ESO) and a Sliding Mode Control (SMC) model, the ESO can effectively observe the speed error and load error of the PMSM, while the SMC model ensures rapid convergence of these errors. It is worth noting that the ADRC system can also incorporate various other types of observers and control models. In recent years, several ADRC strategies have been proposed and studied. For instance, Yang et al. introduced a novel ADRC strategy featuring a double degree-of-freedom (DOF) control model and an extended Kalman filter (EKF). This strategy aims to achieve fast dynamic response and robust anti-interference capability for PMSMs. The effectiveness of this proposed strategy was validated through comprehensive simulations and experimental analyses [8]. In addition, Gabbi et al. proposed a discrete-time SMC model that leverages disturbance observers and discrete digital time delay to mitigate speed fluctuations (caused by load variations) in PMSMs. The simulation and experimental results substantiated the efficacy of this method [9].

However, the insufficiency of accurate PMSM modeling poses a challenge to the effectiveness of the traditional ADRC strategy, particularly in low-speed conditions and for unconventional PMSM structures such as IPMSM, switched reluctance motor (SRM) and multiple-phase PMSM [10–12]. In addition, while the ADRC strategy primarily focuses on speed-loop regulation in PMSM, theoretical analysis indicates that current-loop regulation significantly impacts speed stability and anti-disturbance performance in PMSM [13,14]. Therefore, this study delves into the composite control of speed-loop and current-loop in IPMSM, with a specific emphasis on assessing and comparing three control models for the speed-loop. Notably, the IPMSM distinguishes itself from traditional PMSM through the inclusion of permanent magnets embedded in the rotor, resulting in high power density and a superior torque/inertia ratio [15–17].

This paper appraises a composite control strategy for improving the speed stability and anti-interference performance of IPMSM of electric vehicles. The primary objectives of this research are outlined as follows. Firstly, analysis and presentation of the theoretical model for the composite control strategy of the IPMSM are conducted. The current-loop regulator was implemented as a PI regulator, while the speed-loop regulators consisted of the basic SMC regulator, the linear SMC–ESO regulator and the nonlinear SMC–ESO regulator. Secondly, a comparative analysis of the theoretical simulation results was performed among the speed-loop regulators, namely, the basic SMC, linear SMC–ESO and nonlinear SMC–ESO, using the same parameter settings as the PI regulator. The findings of the theoretical simulation analysis demonstrated that the nonlinear SMC–ESO regulator exhibited the most effective performance in enhancing the anti-interference capabilities of the IPMSM. Thirdly, the theoretical simulation analysis results were validated through hardware experimental testing. The verification results indicated that both the linear SMC–ESO and nonlinear SMC–ESO regulators yielded similar effects in maintaining the anti-interference performance of the IPMSM. Finally, a comprehensive discussion and analysis are conducted to explicate the differences between the theoretical simulation analysis results and the hardware experimental testing results.

## 2. Mathematical Model and Regulator Design of IPMSM

### 2.1. Mathematical Model of IPMSM

The mechanical motion equation of IPMSM can be written as

$$J\frac{d\omega_m}{dt} = T_e - T_L - B\omega_m \tag{1}$$

where $J$ represents the rotational inertia, $\omega_m$ denotes the mechanical angular velocity, $T_e$ indicates the electromagnetic torque, $T_L$ signifies the load torque and $B_m$ expresses the damping coefficient [18]. Additionally, in terms of a dq two-phase synchronous rotating coordinate system, if the d-axis current $i_d$ is assumed as $i_d = 0$, then the electromagnetic torque $T_e$ can be expressed as

$$T_e = \frac{3}{2} P_n \psi_f i_q \tag{2}$$

where $P_n$ represents the number of pole pairs, $\psi_f$ denotes the excitation flux (excitation flux linkage of permanent magnets) and $i_q$ signifies the q-axis current.

### 2.2. Current-Loop Regulator Design of IPMSM

In some references of PMSM and IPMSM, the current-loop was also named as the inner-loop [19]. This paper adopts a PI regulator as the current-loop regulator of IPMSM. In terms of a dq two-phase synchronous rotating coordinate system, the PI regulator of IPMSM can be expressed as

$$\begin{cases} u_d^* = \left(k_{pd} + \frac{k_{id}}{s}\right)\left(i_{dref} - i_{d-}\right) - \omega_e L_q i_{q-} \\ u_q^* = \left(k_{pq} + \frac{k_{iq}}{s}\right)\left(i_{qref} - i_{q-}\right) + \omega_e \left(L_d i + d_- \psi_f\right) \end{cases} \tag{3}$$

where the subscript symbols $d$ and $q$ represent the d-axis and q-axis, respectively, $u_d^*$ and $u_q^*$ denote the voltages, $k_{pd}$ and $k_{pq}$ signify the proportional coefficient, $k_{id}$ and $k_{iq}$ depict the integral coefficient, $i_{dref}$ and $i_{qref}$ express the set value of current, $i_{d-}$ and $i_{q-}$ convey the feedback value of current, $L_d$ and $L_q$ stand for the inductances and $\omega_e$ refers to the electric angular velocity [20].

After theoretical analysis, the regulating parameters of the PI regulator for the d-axis current-loop and q-axis current-loop of IPMSM can be written as

$$\begin{cases} K_{pd} = \alpha L_d \\ K_{id} = \alpha R \\ K_{pq} = \alpha L_q \\ K_{iq} = \alpha R \end{cases} \tag{4}$$

where $\alpha = 2\pi / \min\{L_d/R, L_q/R\}$ and $R$ is the resistance of stator phase winding.

The purpose of the current-loop regulator design was to swiftly provide the desired values for the d-axis current and q-axis current of IPMSM. However, the specific magnitude of the desired values for the d-axis current and q-axis current was determined by the speed-loop regulator of the IPMSM.

### 2.3. Speed-Loop Regulator Design of IPMSM

This paper introduces three speed-loop regulators, namely, the basic SMC regulator, the linear SMC–ESO regulator and the nonlinear SMC–ESO regulator, to investigate the speed stability and anti-interference performance of the IPMSM.

(1) Basic SMC regulator design

In the d-axis current term $i_d = 0$, the q-axis current and mechanical angular speed of the IPMSM can be written as a new equation system [21]:

$$\begin{cases} \frac{di_{qref}}{dt} = \frac{1}{L_q}\left(-R i_{qref} - P_n \psi_f \omega q_m()\right) \\ \frac{d\omega_m}{dt} = \frac{1}{J}\left(-T_L + \frac{3}{2} P_n \psi_f i_{qref}\right) \end{cases} \tag{5}$$

Define parameter $x_1$ as the error between mechanical angular speed' constant value $\omega_{ref}$ and feedback value $\omega_m$:

$$\begin{cases} x_1 = \omega_{ref} - \omega_m \\ \dot{x}_1 = \dot{\omega}_{ref} - \dot{\omega}_m = -\dot{\omega}_m = -\frac{3}{2J} P_n \psi_f i_{qref} \\ \ddot{x}_1 = -\ddot{\omega}_m = -\frac{3}{2J} P_n \psi_f \dot{i}_{qref} \end{cases} \tag{6}$$

where the symbol '•' signifies a derivation. Adopt the basic sliding mode surface and derive its expression as

$$\begin{cases} s = cx_1 - \dot{x}_1 \\ \dot{s} = c\dot{x}_1 - \ddot{x}_1 = -c\frac{3}{2J}P_n\psi_f i_{\text{qref}} + \frac{3}{2J}P_n\psi_f \dot{i}_{\text{qref}} = -\varepsilon sgn(s) - gs \end{cases} \tag{7}$$

Then, the q-axis current in Equation (7) can be expressed as

$$i_{\text{qref}} = \frac{2J}{3P_n\psi_f} \int_0^t \left( c\frac{3}{2J}P_n\psi_f i_{\text{qref}} - \varepsilon sgn(s) - gs \right) dt \tag{8}$$

where $c$ represents the design parameters, $\varepsilon$ denotes the constant velocity approach rate, $sgn(s)$ signifies the symbolic function and $g$ indicates the exponential approach rate.

Equation (8) indicates that the error $x_1$ between mechanical angular speed' constant value $\omega_{ref}$ and feedback value $\omega_m$ can be converted to the q-axis current $i_{\text{qref}}$ and the q-axis current $i_{\text{qref}}$ is the input value of current-loop of IPMSM (see Equation (3)). Therefore, Equation (8) signifies the basic SMC regulator of IPMSM.

(2)    Linear SMC–ESO regulator design

ESO can also be applied in various fields such as mechanical processing, robotics motion control and unmanned aerial vehicle systems, etc. In this study, the load interference and speed fluctuation of the IPMSM were considered as new parameters by the ESO. The load interference refers to external interference, while the speed fluctuation pertains to internal interference.

According to the Equation (1) of IPMSM mathematical model, the linear second-order ESO can be written as

$$\begin{cases} \dot{z}_1 = \dot{z}_2 - 2p(z_1 - \omega) + \frac{3}{2J}P_n\psi_f i_q \\ \dot{z}_2 = -p^2(z_1 - \omega) \end{cases} \tag{9}$$

where $z_1$ represents the speed fluctuation observer of IPMSM, $z_2$ denotes the load interference observer of IPMSM and $p$ indicates the pole of linear second-order ESO [22,23].

Figure 1 shows the basic structure of linear second-order ESO (corresponding to Equation (1)). In Figure 1, the transfer function from $\omega$ to $z_1$ can be written as

$$z_1(s) = \frac{2ps + p^2}{s^2 + 2ps + p^2}\omega(s) + \frac{s}{s^2 + 2ps + p^2}\frac{3}{2J}P_n\psi_f i_q \tag{10}$$

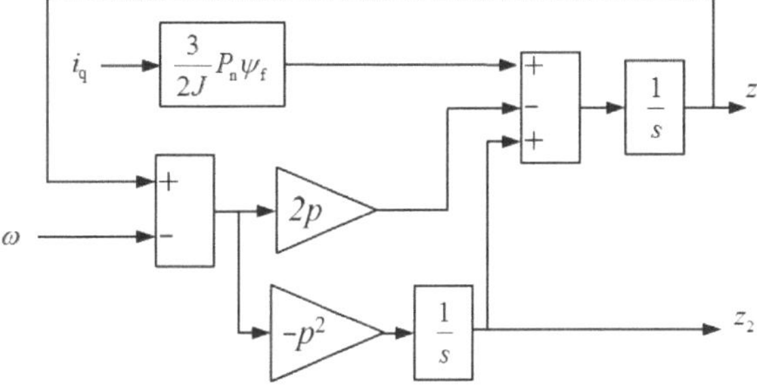


**Figure 1.** The basic structure of linear second-order ESO.

If the q-axis current $i_q = 0$, then Equation (10) can be simplified as

$$z_1(s) = \frac{2ps + p^2}{s^2 + 2ps + p^2}\omega(s) \tag{11}$$

Equation (11) indicates that the speed fluctuation observer $z_1$ is the low pass filter of IPMSM's speed $\omega$.

In the same way, the transfer function from $i_q$ to $z_2$ can be written as

$$z_2(s) = 3P_n\psi_f i_q \frac{p^2}{2J(s^2 + 2ps + p^2)} \tag{12}$$

Equation (12) shows that the load interference observer $z_2$ is the second-order low pass filter of IPMSM's q-axis current $i_q$.

According to the ESO theory, the pole $p$ decides the performance of linear second-order ESO. The more distance between pole $p$ and the imaginary axis, the better anti-disturbance and response performance of load interference observer $z_2$. However, the more distance between pole $p$, the more the imaginary axis can distort the filter effect of speed fluctuation observer $z_1$, which may even result in system instability of IPMSM. Therefore, the value of pole $p$ should be selected based on the actual situation of IPMSM, especially in the condition of imprecise mathematical model of IPMSM.

After making a transformation on the load interference observer $z_2$ of a linear second-order ESO, and subsequently transmitting it to the input variable of the current loop of an IPMSM, a speed-loop regulator for the IPMSM was achieved, which was named as the linear SMC–ESO regulator.

(3)     Nonlinear SMC–ESO regulator design

Based on Equation (9)'s linear second-order ESO, a nonlinear second-order ESO was proposed in this paper, and its mathematical expression can be written as

$$\begin{cases} \dot{z}_1 = \dot{z}_2 - 2p(z_1 - \omega) + \frac{3}{2J}P_n\psi_f i_q \\ \dot{z}_2 = -p^2 \cdot \text{fal}\left[\left(\left(\dot{z}_1 - \dot{\omega}\right) + 2p \cdot (z_1 - \omega)\right), \alpha, \delta\right] \end{cases} \tag{13}$$

where $\alpha$ represents the design parameter, $\delta$ denotes the filter factor and the nonlinear function $\text{fal}((z_1 - \omega), \alpha, \delta)$ can be written as

$$\text{fal}((z_1 - \omega), \alpha, \delta) = \begin{cases} |z_1 - \omega|^\alpha \cdot \text{sign}(z_1 - \omega) & |z_1 - \omega| > \delta \\ \frac{z_1 - \omega}{\delta^{(1-\alpha)}} & |z_1 - \omega| \leq \delta \end{cases} \tag{14}$$

In Equations (13) and (14), the design parameter $\alpha$ represents a constant between 0 and 1. Under normal circumstances, the value range of filter factor $\delta$ is from $5T$ to $10T$, where $T$ denotes the sample time. The larger filter factor $\delta$, the better the filter effect of nonlinear second-order ESO and nonlinear function $\text{fal}((z_1 - \omega), \alpha, \delta)$.

In addition, the smaller design parameter $\alpha$, the better target tracking ability of nonlinear second-order ESO and nonlinear function $\text{fal}((z_1 - \omega), \alpha, \delta)$. However, it should be noted that the smaller design parameter $\alpha$ reduces the filter effect of nonlinear second-order ESO and nonlinear function $\text{fal}((z_1 - \omega), \alpha, \delta)$. Therefore, before the design parameter $\alpha$ is decided, comprehensive experimental testing and analysis should be carried out between the design parameter $\alpha$ and filter factor $\delta$.

By incorporating the Equation (13) form of the nonlinear second-order ESO into the linear SMC–ESO regulator, an enhanced SMC–ESO regulator, referred to as the nonlinear SMC–ESO regulator, was devised.

## 3. Simulation Analysis

To verify the effectiveness of the aforementioned current-loop regulator and speed-loop regulators, the simulation models of the IPMSM were established, and the basic structure of IPMSM was shown in reference [21]. Figure 2 illustrates the system framework of the IPMSM simulation model.

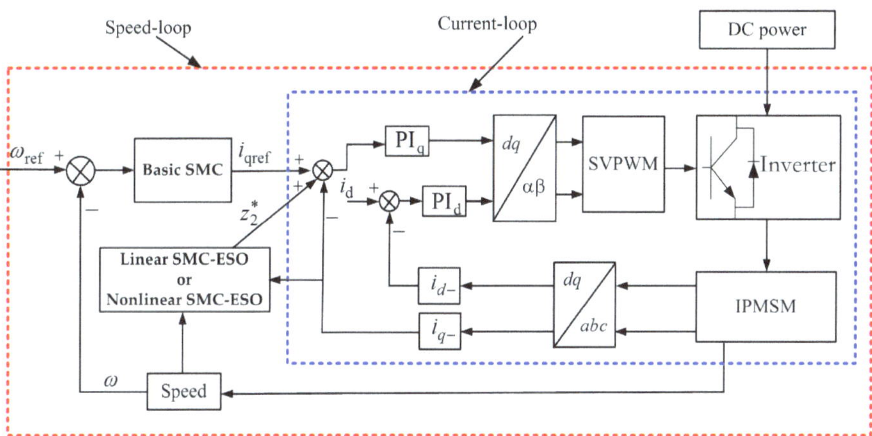


**Figure 2.** System frame of simulation model of IPMSM.

In Figure 2, the d-axis current-loop regulator $\mathrm{PI}_d$ and q-axis current-loop regulator $\mathrm{PI}_q$ of IPMSM were the PI regulator. The speed-loop regulators of IPMSM were the basic SMC regulator, linear SMC–ESO regulator and nonlinear SMC–ESO regulator. During the simulation process, the input value $i_d$ of d-axis current-loop remained zero, and the regulating parameters of d-axis current-loop regulator and q-axis current-loop regulator of IPMSM were decided by the Equation (4), which tested and compared the performance differences among the above three speed-loop regulators.

The main electrical parameters of IPMSM are shown in Table 1.

**Table 1.** Electrical parameters of IPMSM.

| Item | Value | Unit |
|---|---|---|
| Number of phases | 3 | - |
| Number of pole pairs | 4 | - |
| Power | 1.5 | kW |
| Rated voltage | 220 | V |
| Rated current | 4.5 | A |
| Phase resistance | 2.92 | $\Omega$ |
| d-axis inductance | 8.96 | mH |
| q-axis inductance | 12.29 | mH |
| Flux linkage | 0.2388 | Wb |
| Rotor inertia | 0.00104 | kg·m$^2$ |
| Damping coefficient | 0 (assumed value) | N·m·s/rad |

The other basic parameters parameter settings of simulation process of IPMSM are presented as follows:

(1)　d-axis current-loop regulator: $\alpha = 2000$, $K_{\mathrm{pd}} = 17.92$, $K_{\mathrm{id}} = 5840$;

(2)　q-axis current-loop regulator: $\alpha = 2000$, $K_{\mathrm{pq}} = 24.58$, $K_{\mathrm{iq}} = 5840$;

(3)　Speed-loop regulator (basic SMC regulator): $c = 240$, $\varepsilon = 200$, $g = 100$.

In Figure 3, the start-up response of IPMSM in the absence of any external load is depicted. The pole $p$ of ESO was set at 500, while the design parameter $\alpha$ and filter factor $\delta$ of the nonlinear function $\mathrm{fal}((z_1 - \omega), \alpha, \delta)$ were assigned values of 0.001 and 10, respectively. Based on the observations made from Figure 3a, it can be deduced that the overshoot and stabilization time of the speed response of the IPMSM remain nearly identical across the three speed-loop regulators (basic SMC regulator, linear SMC–ESO regulator and nonlinear SMC–ESO regulator). Furthermore, in the steady-state speed of the IPMSM, the steady-state value of the input q-axis current (i.e., the sum of q-axis current $i_{\text{qref}}$ and ESO's observed current $z_2^*$, as shown in Figure 2) obtained from the three speed-loop regulators also exhibited identical characteristics. Therefore, Figure 3 provides evidence that both the linear SMC–ESO regulator and the nonlinear SMC–ESO regulator did not exhibit superiority over the basic SMC regulator during the start-up process and steady-state speed of the IPMSM.

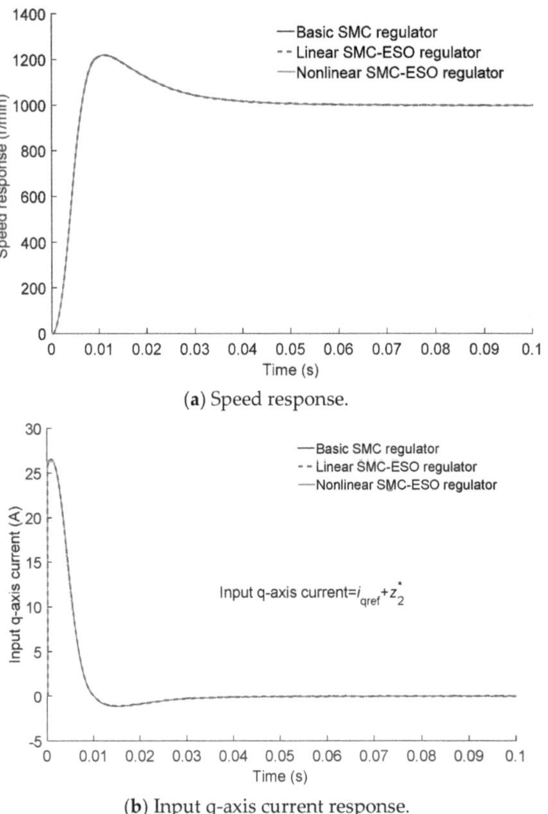


(**a**) Speed response.

(**b**) Input q-axis current response.

**Figure 3.** Start-up response of IPMSM with no-load.

At 0.2 s, a step load of 5 N·m was applied to the rotor of the IPMSM. The subsequent anti-disturbance performance of the IPMSM, including the speed response and input q-axis current response, is depicted in Figure 4. In Figure 4a, the speed of the IPMSM decreases to 891 r/min when employing the fundamental SMC regulator. Conversely, the speed reductions for the linear SMC–ESO regulator and the nonlinear SMC–ESO regulator are 926 r/min and 947 r/min, respectively. Notably, the steady-state speed recovery time of the IPMSM using the nonlinear SMC–ESO regulator was shorter compared to that of both the fundamental SMC regulator and the linear SMC–ESO regulator. Therefore, Figure 4a demonstrates that the nonlinear SMC–ESO regulator minimized the speed fluctuation of the IPMSM and served as the optimal speed-loop regulator for enhancing the anti-disturbance performance of the IPMSM. Figure 4b illustrates the input q-axis current response corresponding to the speed response in Figure 4a. In Figure 4b, the response times for these three regulators were 5.17 ms, 2.72 ms and 0.64 ms, respectively. This phenomenon indicated that the nonlinear SMC–ESO regulator exhibited a faster response and superior anti-disturbance performance. Moreover, the nonlinear SMC–ESO regulator improved the steady-state speed recovery of the IPMSM by minimizing the largest overshoot and secondary fluctuation of the input q-axis current.

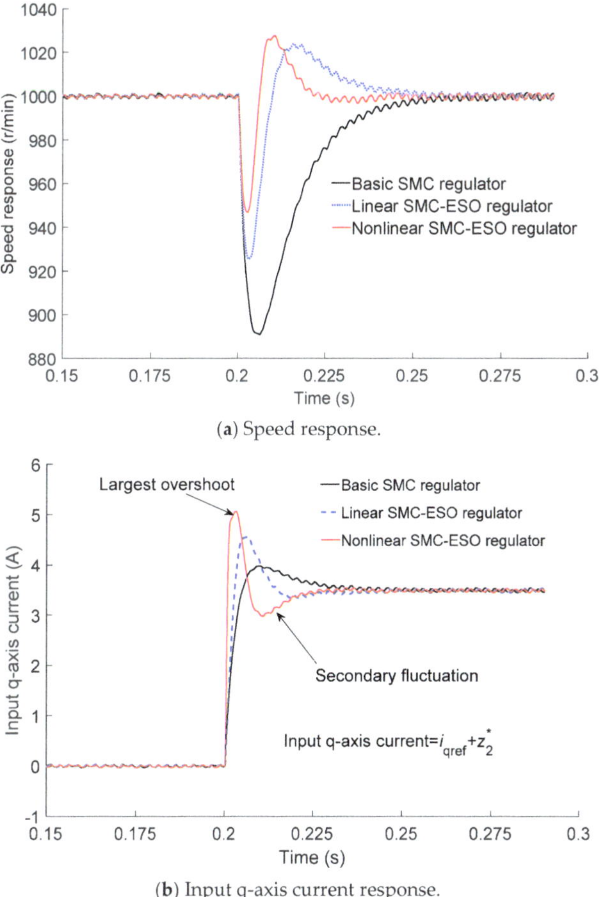



(**a**) Speed response.

(**b**) Input q-axis current response.

**Figure 4.** Anti-disturbance performance of IPMSM with step load.

The q-axis current response of IPMSM under step load conditions is depicted in Figure 5. The step load magnitude was 5 N·m. A comparative analysis between Figures 5 and 4b serves to further substantiate the superior performance and enhanced disturbance rejection capabilities of the nonlinear SMC–ESO regulator.

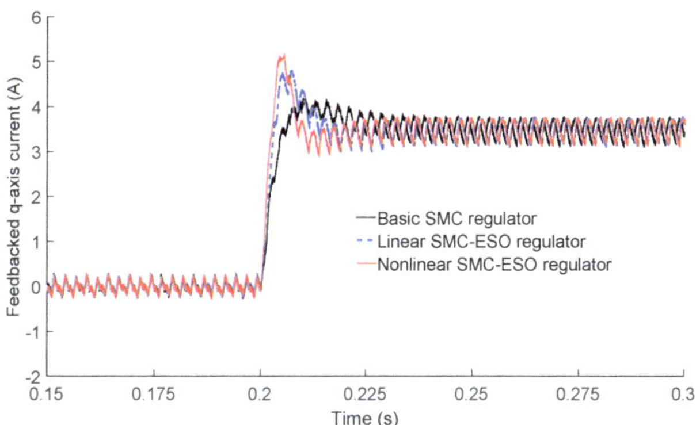



**Figure 5.** Feedback q-axis current of IPMSM step load.

## 4. Experimental Test and Improvement

### 4.1. Experimental Setup

The experimental setup of the IPMSM system is illustrated in Figure 6. It comprised a PC, an IPMSM, an encoder (installed in the back end part of IPMSM), a DC power source, an inverter, a controller, three current sensors, a magnetic particle brake and a power meter. The PC-controller data communication facilitated real-time and visual control of the IPMSM, which was commonly referred to as the real-time control system in the relevant literature. Throughout the experimental testing process, the software program sampled at a frequency of 0.001 s, while the inverter operated at a switching frequency of 1000 Hz. In addition, the electrical parameters of the IPMSM aligned with those specified in Table 1 of the simulation analysis section.

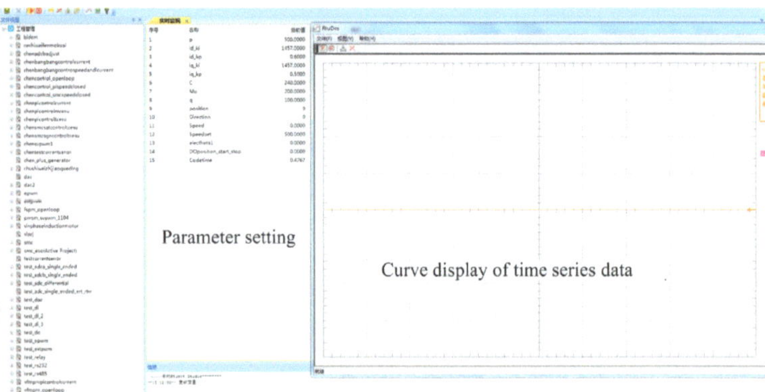


(**a**) PC (software system).

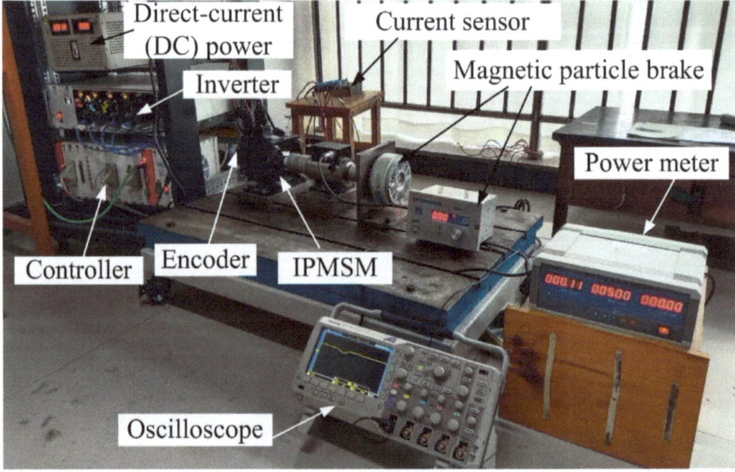


(**b**) Hardware Platform.

**Figure 6.** Experimental setup of the IPMSM system.

### 4.2. Speed Stability Test and Improvement

It should be acknowledged that differences often arise between the theoretical model of the IPMSM and its hardware implementation. These differences are due to the inherent limitations of theoretical models in accurately capturing the intricacies of the hardware system. Factors such as the damping coefficient, rotor inertia, phase resistance and inductance of the IPMSM are subject to variations induced by the winding temperature, rotor speed and load conditions. Therefore, the experimental testing process started up with a focus on assessing and enhancing the speed stability of the IPMSM.

Figure 7 depicts the comparison of speed stability for IPMSM. The parameters for the speed-loop regulator (basic SMC regulator) were maintained as $c - 240$, $\varepsilon = 200$ and $g = 100$. The original Proportional–Integral (PI) parameters and the improved PI parameters for the current-loop regulators are presented in Table 2. The speed fluctuation rate of the IPMSM, calculated using Equation (4) with the original PI parameters from Table 2, amounted to 18.2% ((maximum value − minimum value)/set value). However, considering the switching frequency of the inverter and based on comparative test results, the improved PI parameters were determined through an exploratory method (test results comparison). Accordingly, the speed fluctuation rate of the IPMSM was reduced to 3.8%.

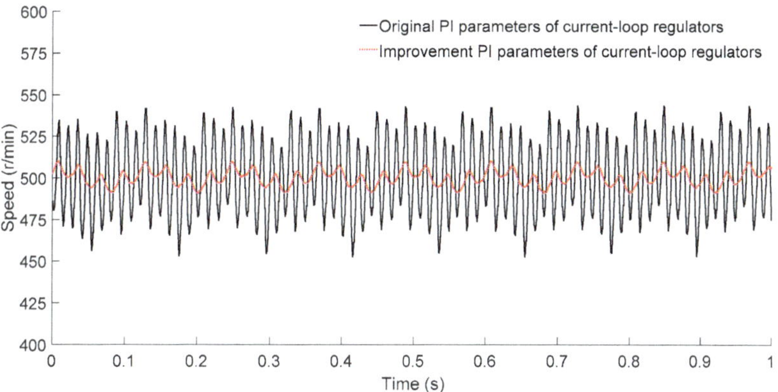


**Figure 7.** Speed stability comparison of IPMSM (measured by the current sensor and collected by the software system of PC, see Figure 6a,b).

**Table 2.** PI parameters of d-axis current-loop regulator and q-axis current-loop regulator.

| Item | Original PI Parameters | Improvement PI Parameters |
| --- | --- | --- |
| d-axis current-loop regulator | $k_{pd} = 17.92, k_{id} = 5840$ | $k_{id} = 5840, k_{id} = 1457$ |
| q-axis current-loop regulator | $k_{pq} = 24.58, k_{iq} = 5840$ | $k_{pq} = 0.5, k_{iq} = 1457$ |

Moreover, to ensure the comparability of speed test results and avoid the influence of high-order harmonics from the oscilloscope, the speed data of the IPMSM/s, as shown in Figure 7, were measured using a current sensor and collected by a PC-based software system. During the experimental test process, the self-protection mechanism of the inverter caused frequent shutdowns when the speed set value of the IPMSM was 1000 r/min. Therefore, a speed set value of 500 r/min was employed for all hardware tests conducted on the IPMSM system.

Moreover, the comparison of speed stability results, based on the linear SMC–ESO regulator and the nonlinear SMC–ESO regulator for the speed-loop regulator, indicated that the improved PI parameters for the current-loop regulators could effectively reduce the speed fluctuation rate of the IPMSM. Therefore, in the subsequent sections concerning the start-up response test and anti-disturbance performance test, the improved PI parameters from Table 2 were adopted for the d-axis current-loop regulator and q-axis current-loop regulator of the IPMSM.

### 4.3. Start-Up Response Test

Premised on the enhanced PI parameters of the d-axis current-loop regulator and q-axis current-loop regulator, the start-up response of IPMSM was assessed and compared using the aforementioned three speed-loop regulators, as depicted in Figure 8. Figure 8 shows that the start-up speed of the IPMSM was measured by the power meter and displayed by the oscilloscope (Figure 6b), and the target constant-speed was 500 r/min.

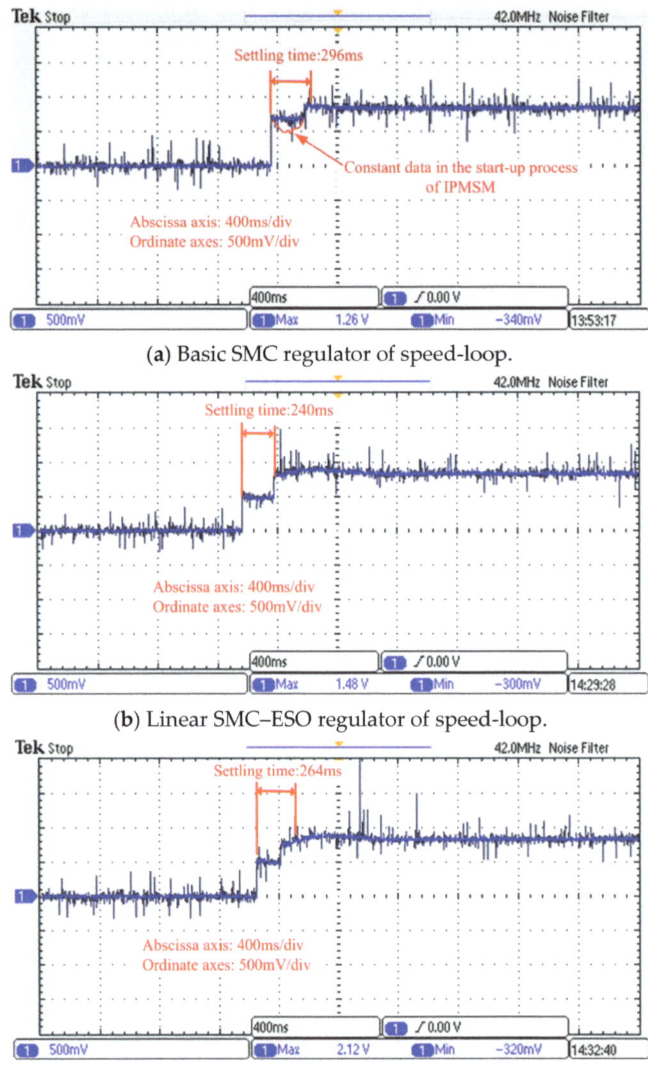



(**a**) Basic SMC regulator of speed-loop.

(**b**) Linear SMC–ESO regulator of speed-loop.

(**c**) Nonlinear SMC–ESO regulator of speed-loop.

**Figure 8.** Start-up response test of IPMSM (measured by the power meter and displayed by the oscilloscope).

For the basic SMC regulator in the speed-loop, the settling time of the IPMSM's start-up speed was measured to be 296 ms. Conversely, for the linear SMC–ESO regulator and the nonlinear SMC–ESO regulator in the speed-loop, the settling times of the IPMSM's start-up speed were determined to be 240 ms and 264 ms, respectively. In summary, Figure 8 demonstrates that the settling time of the IPMSM's start-up speed remained nearly identical across the three aforementioned speed-loop regulators.

Comparing the tested average settling time of the IPMSM with the simulation analysis results presented in Figure 3a, it was evident that the former was smaller. This discrepancy can be attributed to the improved PI parameters of the d-axis current-loop regulator and q-axis current-loop regulator. Additionally, the start-up process of the IPMSM exhibited anomalous constant data phenomena (constant data in the start-up process of IPMSM of Figure 8a) due to the low output resolution of speed measurement for the power meter.

### 4.4. Anti-Disturbance Performance Test

To assess the anti-disturbance performance of the three speed-loop regulators, Figure 9 presents the response of IPMSM to a step load, with a constant speed of 500 r/min and a step load of 1.5 N·m. When the step load was applied to the rotor of the IPMSM, the settling time and speed fluctuation of the basic SMC regulator were approximately 760 ms and 147 r/min, respectively. However, when considering the linear SMC–ESO regulator and the nonlinear SMC–ESO regulator, the settling times were approximately 368 ms and 320 ms, respectively, with corresponding speed reductions of 94 r/min and 88 r/min. Therefore, it was evident that the anti-disturbance performances of both the linear SMC–ESO regulator and the nonlinear SMC–ESO regulator surpass that of the basic SMC regulator.

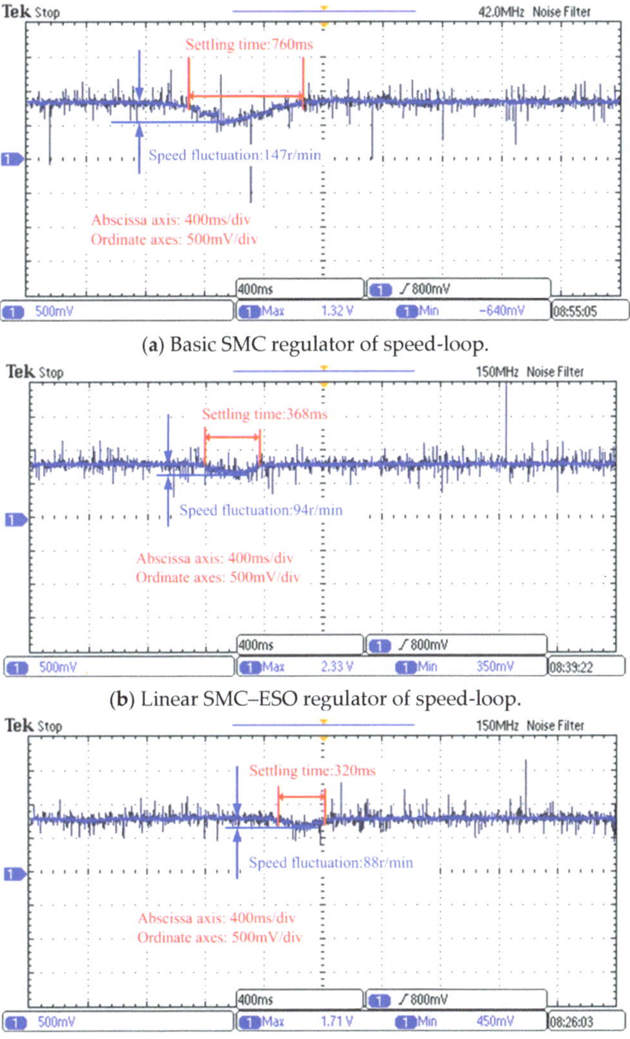


(**a**) Basic SMC regulator of speed-loop.



(**b**) Linear SMC–ESO regulator of speed-loop.



(**c**) Nonlinear SMC–ESO regulator of speed-loop.

**Figure 9.** Anti-disturbance performance test of IPMSM.

Additionally, throughout the multiple tests conducted to evaluate the anti-disturbance performance of the IPMSM, the disparity in settling time and speed fluctuation between the linear SMC–ESO regulator and the nonlinear SMC–ESO regulator was not pronounced (in comparison to the simulation results depicted in Figure 4a). This phenomenon can be attributed to the following factors:

(1)　The sampling frequency of the utilized software program was relatively low, resulting in a slower and less accurate transmission of the speed signal from the encoder to the controller;

(2)　The output resolution of speed measurement for the power meter was insufficient, thereby impeding the timely collection of the differences in settling time and speed fluctuation between the linear SMC–ESO regulator and the nonlinear SMC–ESO regulator by the oscilloscope.

## 5. Discussion

In the speed stability test, the speed fluctuation rate of IPMSM was reduced from 18.2% to 3.8% through the implementation of improved PI parameters in the current-loop regulator (Figure 7 of Section 4.2). However, when the same improved PI parameters of the current-loop regulator in the hardware system were applied to the simulation model of the IPMSM and the parameters of the speed-loop regulator (basic SMC regulator) were kept consistent with those of the hardware system, the speed fluctuation rate of the simulation results for the IPMSM increased, as depicted in Figure 10.

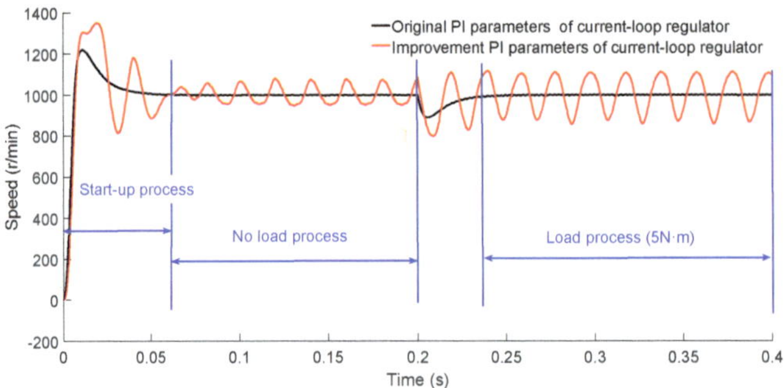


**Figure 10.** The simulation results of current-loop regulator with original and improved PI parameters.

From the analysis presented in Figure 10, it can be deduced that the implementation of the improved PI parameters in the current-loop regulator led to a higher-speed fluctuation rate for the IPMSM throughout its operational processes. Specifically, the settling time of the start-up process was approximately 0.6 s, the speed fluctuation rate during the no-load process was approximately 13.2% and the speed fluctuation rate during the load process was approximately 25.3%. The corresponding reasons are as follows:

(1)　Discrepancies arose between the damping coefficient employed in the hardware testing phase and that utilized in the simulation analysis. In the former, the enhanced PI parameters were derived based on the actual damping coefficient of IPMSM. However, due to the challenge of evaluating the damping coefficient of IPMSM, it was assumed to be zero during the simulation analysis;

(2)　The simulation model of IPMSM did not precisely correspond to the features of its hardware system, resulting in inaccuracies pertaining to the flux linkage, rotor inertia and dq-axis inductances;

(3)　Throughout the hardware testing phase, the actual power quality of the inverter, the accuracy of the encoder and current sensors and the execution process of the software program exerted a significant influence on the selection of optimal PI parameters.

Therefore, the comparison between the results obtained from simulation analysis and hardware testing may provide valuable insights for the practical application of IPMSM in electric vehicle drive systems.

## 6. Conclusions

This paper adopts three speed-loop regulators to investigate the stability and anti-interference performance of IPMSM. Firstly, following theoretical analysis and modeling, the simulation results indicated that the three speed-loop regulators possessed the same functionality in ensuring the speed stability of IPMSM. However, the nonlinear SMC–ESO regulator emerged as the optimal speed-loop regulator for enhancing the anti-interference performance of IPMSM. Secondly, based on the simulation results, the PI parameters of the current-loop regulator of IPMSM were refined during the experimental testing phase, with the objective of improving the speed stability of the IPMSM system. Thirdly, leveraging the improved PI parameters of the current-loop regulator, the anti-interference performance of IPMSM was assessed. The test results demonstrated that both the linear SMC–ESO regulator and the nonlinear SMC–ESO regulator outperformed the basic SMC regulator. However, due to limitations inherent in the software and hardware systems, such as the sampling frequency, power meter and actual physical model, the linear SMC–ESO and nonlinear SMC–ESO regulators exhibited similar effects in maintaining the anti-interference performance of IPMSM, which deviated from the simulation results. Finally, the discussion section of this paper analyzes the difference between the simulation results and the hardware experimental testing results. The simulation calculations, hardware experimental tests and difference analysis of the speed-loop regulators may contribute to the practical application of IPMSM in electric vehicle drive systems.

**Author Contributions:** Conceptualization, Z.C. and X.D.; methodology, Z.C.; software, X.D. and M.F.; validation, Z.C.; writing—original draft preparation, Z.C.; writing—review and editing, X.D. All authors have read and agreed to the published version of the manuscript.

**Funding:** This work was financially supported by the Scientific and Technological Project in Henan Province under Grant No. 232102241024.

**Institutional Review Board Statement:** Not applicable.

**Informed Consent Statement:** Not applicable.

**Data Availability Statement:** Data are contained within the article.

**Conflicts of Interest:** The authors declare no conflict of interest.

# References

1. Badini, S.S.; Verma, V. Parameter independent speed estimation technique for PMSM drive in electric vehicle. *Int. Trans. Electr. Energy Syst.* **2021**, *31*, 13071. [CrossRef]
2. Sepulchre, L.; Fadel, M.; Pietrzak-David, M.; Porte, G. MTPV flux-weakening strategy for PMSM high speed drive. *IEEE Trans. Ind. Appl.* **2018**, *54*, 6081–6089. [CrossRef]
3. Wang, J.; Wang, X.; Luo, Z.; Assadian, F. Active disturbance rejection control of differential drive assist steering for electric vehicles. *Energies* **2020**, *13*, 2647. [CrossRef]
4. Fan, S.; Zhang, Y.; Jin, J.; Wang, X.; Tong, C. Deadbeat predictive current control of PMSM drives with an adaptive flux-weakening controller. *IET Power Electron.* **2022**, *15*, 653–753. [CrossRef]
5. Krishnamoorthy, S.; Sanjeevikumar, P.; Holm-Nielsen, J.B. Torque ripple minimization of PMSM using an adaptive Elman neural network-controlled feedback linearization-based direct torque control strategy. *Int. Trans. Electr. Energy Syst.* **2021**, *31*, 12685.
6. Gmati, B.; Jlassi, I.; Khojet El Khil, S.; Marques Cardoso, A.J. Open-switch fault diagnosis in voltage source inverters of PMSM drives using predictive current errors and fuzzy logic approach. *IET Power Electron.* **2021**, *14*, 1059–1072. [CrossRef]
7. Zhou, Z.; Xia, C.; Shi, T.; Geng, Q. Model predictive direct duty-cycle control for PMSM drive systems with variable control-Set. *IEEE Trans. Ind. Electron.* **2021**, *68*, 2976–2987. [CrossRef]
8. Yang, Z.; Yan, Z.; Lu, Y.; Wang, W.; Yu, L.; Geng, Y. Double DOF strategy for continuous-wave pulse generator based on extended Kalman Filter and adaptive linear active disturbance rejection control. *IEEE Trans. Power Electron.* **2022**, *37*, 1382–1393. [CrossRef]
9. Gabbi, T.S.; Hilton, A.G.; Vieira, R.P. Discrete-time sliding mode control based on disturbance observer applied to current control of permanent magnet synchronous motor. *IET Power Electron.* **2021**, *14*, 875–887. [CrossRef]

10. Wang, P.; Xu, Y.; Ding, R.; Liu, W.; Shu, S.; Yang, X. Multi-Kernel neural network sliding mode control for permanent magnet linear synchronous motors. *IEEE Access* **2021**, *9*, 57385–57392. [CrossRef]
11. Apte, A.; Thakar, U.; Joshi, V. Disturbance observer based speed control of PMSM using fractional order PI controller. *IEEE/CAA J. Autom. Sin.* **2019**, *6*, 316–326. [CrossRef]
12. Lamichhane, A.; Zhou, L.; Yao, G.; Luqman, M. Modeling, control and power management of six-phase PMSM based shipboard MVDC distribution System. *Energies* **2020**, *13*, 4229. [CrossRef]
13. Nakayama, Y.D.S. Design of band elimination filter for PMSM vector control operable in overmodulation region of inverter. *Electr. Eng. Jpn.* **2019**, *207*, 45–56. [CrossRef]
14. Lascu, C.; Andreescu, G.D. PLL position and speed observer with integrated current observer for sensorless PMSM drives. *IEEE Trans. Ind. Electron.* **2020**, *67*, 5990–5999. [CrossRef]
15. Makaino, Y.; Doki, S. Position sensorless control for IPMSM of spatial inductance distribution with nonsinusoidal inductance waveform based on pattern matching method. *Electr. Eng. Jpn.* **2017**, *201*, 35–42. [CrossRef]
16. Liu, J.; Gong, C.; Han, Z.; Yu, H. IPMSM model predictive control in flux-weakening operation using an improved algorithm. *IEEE Trans. Ind. Electron.* **2018**, *65*, 9378–9387. [CrossRef]
17. Uddin, M.N.; Rahman, M.M. Online torque-flux estimation-based nonlinear torque and flux control scheme of IPMSM drive for reduced torque ripples. *IEEE Trans. Power Electron.* **2019**, *34*, 636–645. [CrossRef]
18. Schmid, H.C.; Schroedl, M.; Heyden, M.V. Design space analysis including experimental verification for an electrical machine based on a parametric and functional IPMSM model. *IEEE Trans. Ind. Electron.* **2021**, *68*, 7863–7873. [CrossRef]
19. Chen, B.; Shen, A.; Li, P.; Luo, X.; Xu, S.; Xu, J. Restart strategy for sensorless PMSM drive system based on zero-voltage vector. *IET Electr. Power Appl.* **2020**, *14*, 2362–2369. [CrossRef]
20. Yuan, L.; Hu, B.X.; Wei, K.Y.; Chen, S. *The Control Principle and MATLAB Simulation of Modern Permanent Magnet Synchronous Motor*; Beijing University of Aeronautics and Astronautics Press: Beijing, China, 2016.
21. Wei, Y.; Sun, L.; Chen, Z. An improved sliding mode control method to increase the speed stability of permanent magnet synchronous motors. *Energies* **2022**, *15*, 6313. [CrossRef]
22. Bouguenna, I.F.; Azaiz, A.; Tahour, A.; Larbaoui, A. Robust neuro-fuzzy sliding mode control with extended state observer for an electric drive system. *Energy* **2019**, *169*, 1054–1063. [CrossRef]
23. Humaidi, A.J.; Ibraheem, I.K. Speed control of permanent magnet DC motor with friction and measurement noise using novel nonlinear extended state observer-based anti-disturbance control. *Energies* **2019**, *12*, 1651. [CrossRef]

*World Electric Vehicle Journal*

*Article*

# Design and Analysis of a Permanent Magnet Brushless DC Motor in an Automotive Cooling System

Kai Ren [1,*], Hongxuan Chen [1], Haiyang Sun [2], Qin Wang [3,4], Qingyun Sun [1] and Bo Jin [5]

[1] School of Mechanical and Electronic Engineering, Nanjing Forestry University, Nanjing 210037, China
[2] School of Information Science and Technology, Nanjing Forestry University, Nanjing 210037, China
[3] Key Laboratory of Energy Thermal Conversion and Control of Ministry of Education, School of Energy and Environment, Southeast University, Nanjing 210096, China
[4] Jiangsu Jiaoke Energy Technology Development Co., Ltd., Nanjing 210017, China
[5] Huzhou Yueqiu Motor Co., Ltd., Huzhou 313000, China; jin.bo@yqmotor.com
* Correspondence: kairen@njfu.edu.cn

**Abstract:** Conducting excellent thermal management of a new electric vehicle motor drive system may enhance the operational efficiency of the motor drive and minimize its pollutant emissions and energy losses. As an important part of the motor thermal management system, it is necessary to improve the design of the drive motor for the fan. This paper presents the design of a 12s-10p permanent magnet brushless DC motor with a rated speed of 2200 rpm and a rated voltage of 12 V based on finite element analysis. At this rated speed, the maximum torque the motor can output is 1.80 N·m. Then, we calculated the loading capacity of the motor by parameterizing the resistance in the circuit. We have built a prototype based on the design results and built a test bench to test the loading capacity of the prototype. A comparison revealed that the error between the experimental and calculated results was small. Accordingly, it is believed that this work is capable of serving as a theoretical guide for the design and manufacture of automotive cooling fans in the future.

**Keywords:** brushless DC motor; torque; electromagnetic properties; simulations; bench test

**Citation:** Ren, K.; Chen, H.; Sun, H.; Wang, Q.; Sun, Q.; Jin, B. Design and Analysis of a Permanent Magnet Brushless DC Motor in an Automotive Cooling System. *World Electr. Veh. J.* **2023**, *14*, 228. https://doi.org/10.3390/wevj14080228

Academic Editors: Joeri Van Mierlo and Zhongxian Chen

Received: 13 July 2023
Revised: 3 August 2023
Accepted: 16 August 2023
Published: 18 August 2023

## 1. Introduction

Oil consumption has skyrocketed alongside the world's population and the pace of economic development, putting a strain on energy and the environment worldwide [1–5]. This global problem has led to a greater emphasis on the design of new electric vehicles (NEVs) that are eco-friendlier and more efficient [6]. Researchers' relentless pursuit of permanent magnet materials and alternative energy sources has been rewarded with significant theoretical and technological advances. They have ushered in a new era of NEVs, in which electricity has gradually become the predominant power source for automobiles [7], as well as a future which is compliant with energy conservation requirements and in which pollutant emissions and nonrenewable energy consumption are drastically reduced [8].

As motor drive systems are a primary component of NEVs, their performance has a significant impact on NEVs' performance. A motor drive system will continue to accumulate heat during operation [9], and when the system temperature reaches a certain threshold, the system will be exposed to an increasingly hostile environment, wherein the electromagnetic characteristics of the material used in the motor are altered; the residual flux density and coercivity are reduced [10]; the torque capability of the motor is restricted [11]; the power loss is increased [12]; the system's stability and reliability are decreased [13]; and the performance of the drive system is severely limited [14]. In order to prevent these adverse effects, the performance of the cooling system in the motor drive system becomes a particular issue to address.

In an automotive cooling system, the fan is a crucial component. In conventional fuel-powered vehicles, the fan is mechanically driven by a belt connected to the engine crankshaft [15], resulting in limited speed control and inadequate energy efficiency [16]. In contrast, cooling fans in electric vehicles are driven directly by the electric motor, which provides better control and energy efficiency, reduces auxiliary energy consumption, and enables better thermal management of the drive motor [17]. As the drive motors for fans, brushless DC (BLDC) motors are a common choice [18]. By substituting physical commutators with electronic commutators, BLDC motors have successfully avoided the issue of mechanical abrasion of the built-in commutators, which is typical in conventional brushed DC motors [19]. Thus, they benefit from the advantages of reduced costs, lower noise, a simpler structure, less maintenance effort, a longer life [20], more precise speed control, greater efficiency [21], etc.

To further enhance the performance of BLDC motors, numerous researchers have conducted extensive research. Jeon Kyunghun et al. [22] conducted a reliability-based robust design optimization (RBRDO) of BLDC motors utilizing a reliability analysis in an effort to maximize output torque. V Bharanigha et al. [23] proposed a multifunctional APID controller for BLDC motor speed regulation to reduce the generation of motor torque pulsation. Smóka Krzysztof et al. [24] proposed several different design solutions for small permanent magnet brushless DC (PMBLDC) motors with a rotor in the form of a single ferrite magnet, and designs with a lower cogging torque are proposed in comparison. Toren Murat [25] analyzed the performance parameters of permanent magnets at different levels of power density, different operating temperature ranges, and different thicknesses, as well as their influence on motor performance, and determined the optimum values ranges. Saed Nejat et al. [26] compared the noise and vibration characteristics of single-phase BLDC machines with convex and claw pole stators. It was demonstrated that machines with a claw pole stator produced more structural and airborne noise. As can be seen from the above research, although research on motors is more detailed, there is relatively little research on the complete design of a fan drive motor with output powers between 400 W and 500 W. In order to conduct a more detailed study of the motor design at this output power, as well as a more detailed analysis of the application in practice, this paper designs and calculates a cooling fan drive motor in the cooling system of an electric vehicle drive motor with output powers between 400 W and 500 W, simulates its load capacity by parametrizing the resistance in the circuit, and conducts experiments.

In this paper, we propose a design for a PMBLDC motor in an automotive cooling system and apply a finite element analysis to calculate its output characteristics. In addition, the load capacity of the motor at the rated speed is addressed by parameterizing the control circuit resistance, while the output characteristics of the motor are determined with a prototype installed on a specially developed bench test platform. The outcome of the bench test demonstrates the desired accuracy of the simulation analysis.

## 2. Design and Simulation Analysis

The design solutions for a motor that serves as the power source for a cooling fan are comprehensively determined by parameters pertaining to the fan's various operating conditions and applications, such as electromagnetic load, torque, speed, and efficiency. Meanwhile, motor volume, temperature rise, and other variables that can affect the performance of such a fan motor within an automobile engine should be considered.

### 2.1. Main Parameters of the Design

The maximum electromagnetic torque under a continuous duty has a direct positive correlation with the size of a PMBLDC motor with an outer rotor:

$$T_{\text{peak}} \propto (D_a^2 L_{\text{ef}}) \tag{1}$$

where $T_{\text{peak}}$ represents the maximum electromagnetic torque under a continuous duty; $D_a$ is the outer diameter of the stator core, m; and $L_{\text{ef}}$ denotes the calculated length of the stator core, m.

As the two most important motor design parameters, $D_a$ and $L_{ef}$ are determined as follows [27]:

$$D_a^2 L_{ef} = \frac{6.1 \times 10^3 \times P'}{u_i K_{Nm} K_{dp} A B_\delta n} \tag{2}$$

where $P'$ is the calculated power, Kw; $\alpha_i$ is the magnet pole-arc ratio; $K_{Nm}$ is the armature winding coefficient; $K_{dp}$ is the air gap magnetic flux coefficient; $A$ is the armature line load, A/m; $B_\delta$ is the amplitude of air gap magnetic density, T; and $n$ is the rated speed, r/min.

In this study, the outer rotor PMBLDC is designed for a cooling fan; as the design of the PMBLDC has been thoroughly researched, cost reduction became the focus of this design in order to cope with the fierce competition in the market. We used 50AW800 and ferrite as the main materials for our motors. The saturation flux density value of 50AW800 is 2.0 T. At 60 °C, the coercivity value of ferrite is 281.5 kA/m and the remanence value is 0.3746 T. In the design of the motor, the armature line load of the motor was selected within a certain range based on experience, and since the design was based on the maximum output torque of the motor, the value of the armature line load was increased to 23 A/mm. Considering the feasibility of installation, the outer diameter of the motor shell was chosen to be 108.5 mm, the rotor outer diameter was designed to be 93.6 mm, the outer diameter of the stator core was designed to be 91 mm, and the calculated length of the stator core was designed to be 31.2 mm.

The air gap shear stress can be expressed as [28]

$$\sigma = \frac{2T}{\pi D_a^2 L_{ef}} \tag{3}$$

where $T$ is the maximum output torque under a continuous duty with a rated speed.

After calculation, the motor air gap shear stress is 0.70 psi; the motor is not a high torque density motor. In order to reduce costs, the torque density of the motor is not high and the permanent magnet material is ferrite, which has a relatively low remanence, also leading to a motor that does not have a high torque density.

Regarding the number of pole pairs of the motor, a design with a large number of pole pairs will result in an increase in electrical frequency and, consequently, iron loss in the stator. Therefore, the number of motor pole pairs should not be excessive. Motors with a similar number of poles and slots have a higher efficiency and power density than other types of motor. The winding ends are shorter and the torque pulsation is better [29]. In addition, the design adopts a fractional slot winding configuration. In general, the following combination was determined: the number of slots $Z = 12$ and the number of pole pairs $p = 5$. Moreover, two-layer concentrated windings were implemented to further shorten the end windings of the motor and reduce motor losses [30]. Using ANSYS Maxwell, the simulation model for the design of the PMBLDC motor was established based on the above design considerations. Figure 1 depicts the structure and winding configuration of the motor, while Table 1 depicts the other design parameters.

**Table 1.** Parameters of the proposed PMBLDC motor in an electric vehicle cooling fan.

| Parameters. | Value | Parameters | Value |
|---|---|---|---|
| Stator outer diameter | 91 mm | Rated speed | 2200 rpm |
| Stator inner diameter | 28.3 mm | Coil pitch | 1 |
| Rotor outer diameter | 108.4 mm | Number of strands | 1 |
| Rotor inner diameter | 93.6 mm | Winding layers | 2 |
| Stator length | 18 mm | Parallel branches | 1 |
| Rotor length | 31.2 mm | Conductors per slot | 22 |
| Magnet thickness | 5.2 mm | Magnet material | Ferrite (BAOSTEEL) |
| Winding type | Whole-coiled | Steel material | 50AW800 (MASTEEL) |
| Nominal voltage | 12 V | Output power | 415 W |

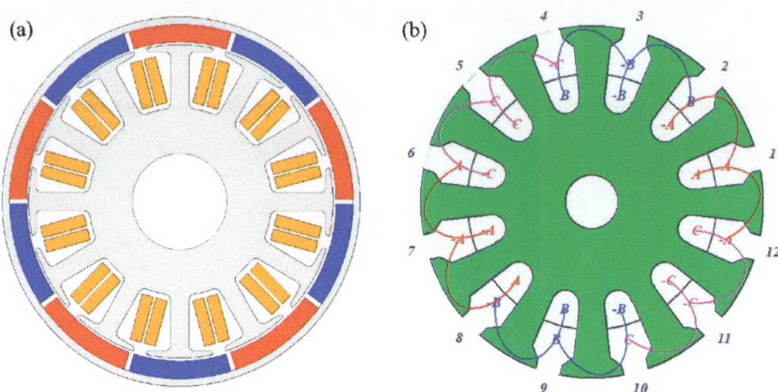

**Figure 1.** The structure of the proposed PMBLDC motor in the automotive cooling system: (**a**) sectional view and (**b**) the winding configuration.

### 2.2. Electromagnetic Field Finite Element Simulation

### 2.2.1. Characteristic Simulation in the No-Load Condition

Based on the above motor structure and parameters, no-load simulations of the motor were carried out on the ANSYS Maxwell platform to verify the validity of the design. Figure 2a,b depicts the magnetic flux density and magnetic lines of force of the motor when it is rotating at an angle of 55° under no-load conditions. Observations reveal that only the stator tooth tip is saturated, and the saturated area is periodically symmetric, which has little impact on the operation of the motor. Moreover, the magnetic density of the stator yoke is not saturated and is not susceptible to demagnetization. Furthermore, the magnetic lines of force are symmetrically distributed, allowing the majority of them to enter the stator without obstruction. In other words, there is almost no magnetic flux leakage, and the distribution of magnetic lines of force is relatively optimal. Thus, it can be concluded that the magnetic flux density and magnetic lines of force of the stator in the no-load condition satisfy the design requirements.

Likewise, the motor was simulated to characterize its cogging torque and radial air-gap flux density under the condition that its external excitation is unloaded; the time step of the simulation was 0.1 s. As depicted in Figure 3a,b, the cogging torque is low, and its relatively ideal waveform is conducive to the smooth operation of the motor and reductions in vibration and noise. On the other hand, the air gap flux density waveform is approximately trapezoidal, which indicates a high efficiency irrespective of the fact that the large fluctuations at the peak and in the numerous harmonic components may reduce the ability of the motor to regulate vibration and noise. Meanwhile, a FFT spectral analysis of the radial air gap flux density (Figure 3c) reveals that among the amplitudes of the fundamental wave, the third harmonic, the fifth harmonic, and the seventh harmonic are 0.5903 T, 0.1762 T, 0.0881 T, and 0.0528 T, respectively, with these three harmonics being the primary contributors to the air gap flux density distortion. Since an increase in harmonic components will increase torque ripple, the motor can be optimized further by decreasing the harmonic content of the air gap flux density.

### 2.2.2. Characteristic Simulation in the Loaded Condition

Following the no-load simulation, the motor was simulated again to characterize its performance under load. The objective was to determine if the design is capable of achieving the expected working performance at a rated supply voltage of 12 V. The control circuit diagram is shown in Figure 4a; the time step of the simulation is 0.0002 s. First, the start-up process of the motor was simulated with a constant output power of 415 W when the mechanical characteristics are considered. Figure 4b depicts the electromagnetic torque waveform of the process from the start-up state to the steady working state of the motor.

A constant value of the electromagnetic torque is observed at approximately 2.14 N·m, suggesting that the output torque can satisfy the actual work requirements. Figure 4c illustrates the revolution speed of the motor during its start-up and steady operation states. The speed rises rapidly at first and stabilizes at around 2200 rpm when reaching the steady state, which also satisfies the requirements. Figure 4d shows the torque-speed curve of the motor during start-up.

Figure 5a,b represents the waveforms of the electromagnetic torque and line current of the motor at 2200 rpm, excluding the mechanical characteristics of the motor. The time step of the simulation is 0.0002 s. The operation of the motor was unstable during the start-up phase. After a period of time, the torque gradually increased from 0 to a constant value. The average torque during the process is approximately 1.80 N·m. The RMS values of the line current is 49.13 A, indicating that the electromagnetic torque in the steady working phase can satisfy the requirements for driving a cooling fan. The simulation data of 1.8 N·m represent the maximum torque that the motor can deliver under these simulation conditions, i.e., the limit of the motor's operation, and does not mean that the motor needs to run at this operation for a long time. Figure 5c depicts the three-phase flux linkage diagram of the motor under rated conditions, which conforms roughly to a sinusoidal distribution, indicating that the electromagnetic properties are stable and the harmonic content is low. Figure 5d shows the iron loss of the motor running in the rated state; the average iron loss is stable at about 33 W, meeting the requirements.

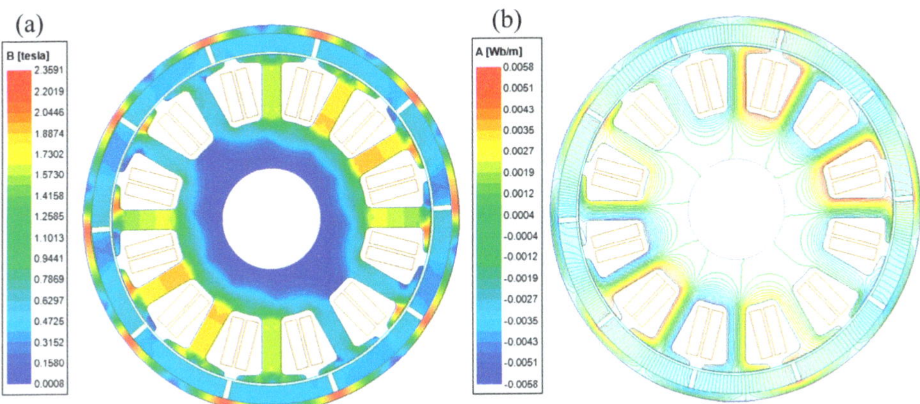

**Figure 2.** (**a**) The magnetic flux density, and (**b**) the magnetic lines of force of the proposed PMBLDC motor model in the no-load condition.

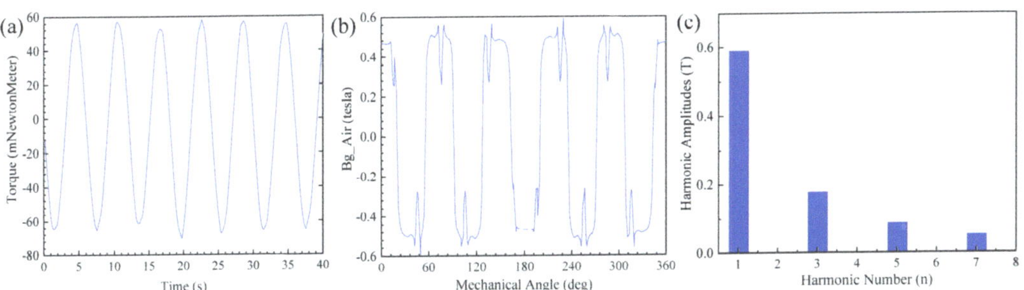

**Figure 3.** (**a**) Simulated cogging torque waveform, (**b**) simulated air gap flux density waveform, and (**c**) FFT result of the air gap flux density of the proposed PMBLDC motor.

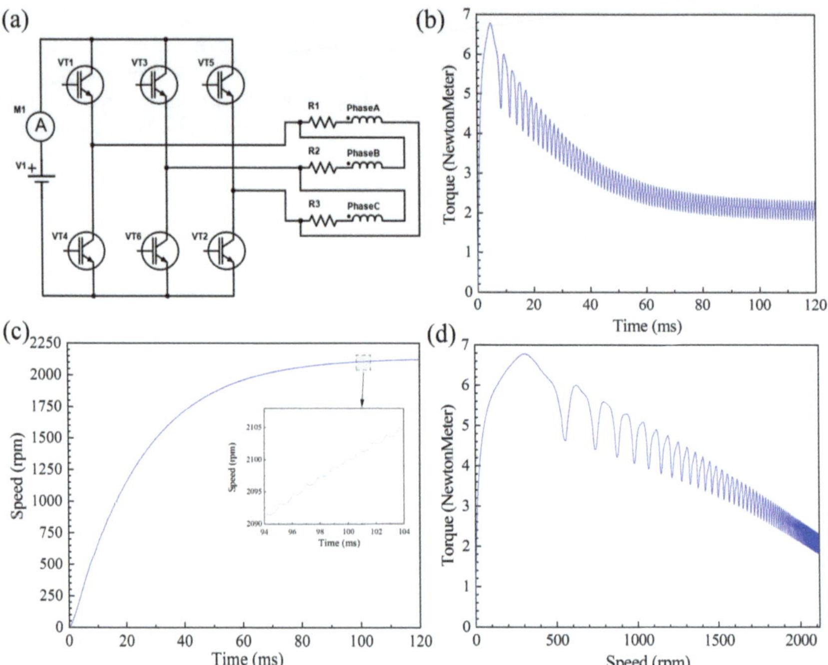

**Figure 4.** (**a**) The control circuit, (**b**) simulated start-up torque waveform, (**c**) simulated start-up speed waveform of the proposed PMBLDC motor, and (**d**) simulated start-up torque-speed waveform.

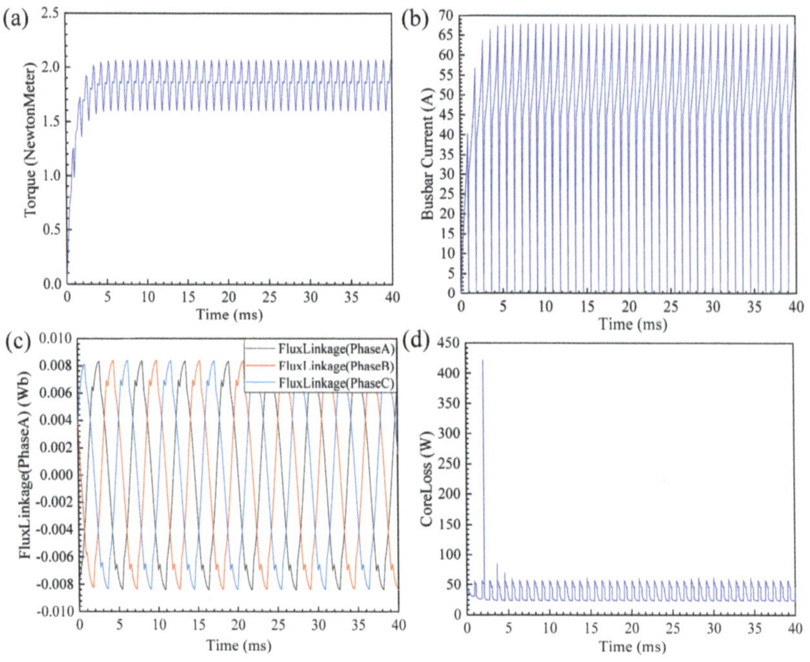

**Figure 5.** (**a**) The calculated electromagnetic torque waveform, (**b**) the Busbar current waveform, (**c**) the three-phase flux linkage diagram of the proposed PMBLDC motor, and (**d**) the iron loss waveform.

## 3. Load Capacity Simulation

A new simulation test was conducted to further confirm the load capacity of the motor. To observe the changes in input current, the external load of the motor was gradually increased at its rated speed (2200 rpm). Since it is difficult to change the external load at a fixed speed in ANSYS Maxwell, a method of reverse verification was adopted: the maximum torque was determined by analyzing the output torque as it varies in response to a change in input current.

Given that the research object is a BLDC motor, the only strategy to add excitation in ANSYS Maxwell is applying an external circuit. In order to alter the input current, various resistances were added to the external circuit resistance throughout this simulation. Among the various schemes for adding the resistance, positions before the motor windings and after the three-phase inverter circuit were designated. A current meter was also included to measure the input current. The complete control circuit diagram is shown in Figure 6a.

After the control circuit is added to the brushless DC motor, the resistance value of the resistance R4 R5 R6 was set to the same parameter, and this parameter was scanned after setting the initial value, final value, and step size. The time step of the simulation was 0.0002 s. Since the power supply voltage remains unchanged, each resistance value corresponds to a current value. This method can be used to simulate multiple different input current values when the motor is running at the rated speed. As depicted in Figure 6b,c, the output electromagnetic torque and output line current vary with resistance. In conjunction with the simulation results obtained when the resistor was disconnected from the control circuit, the maximum electromagnetic torque of the proposed PMBLDC motor can be calculated as 1.80 N·m at an input line current of 49.13 A. This method provides a reference for other applications requiring a change in output torque at a constant speed; by varying the resistance of the control circuit to regulate the input current under a fixed supply voltage, the output electromagnetic torque of the motor can be controlled indirectly.

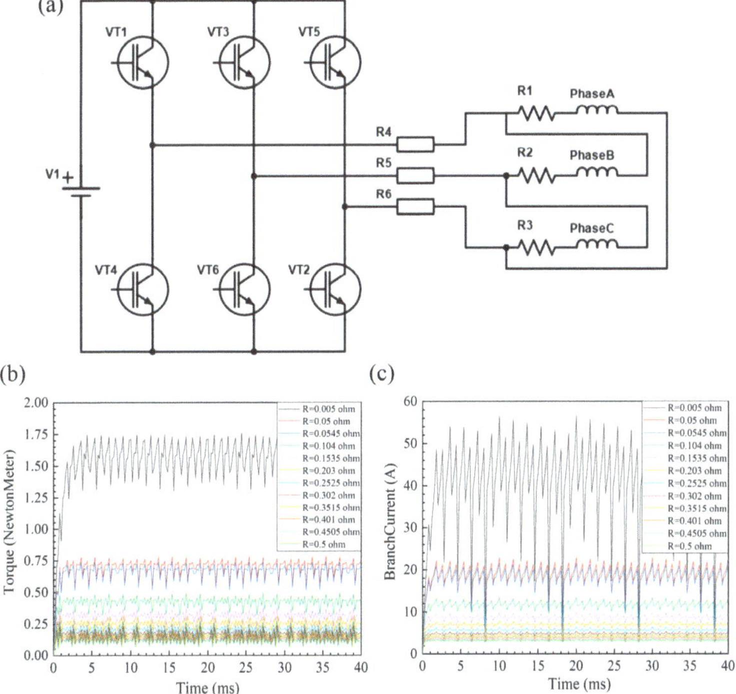


**Figure 6.** (**a**) Variable current control circuit diagram, (**b**) electromagnetic torque at different resistances, and (**c**) line current at different resistances of the proposed PMBLDC motor.

## 4. Prototype Manufacturing and Bench Test

In order to further explore the actual output characteristics of the proposed PMBLDC motor, a prototype was developed and a bench test was performed. The prototype is shown in Figure 7a. The test platform consists of a power tester, tooling, a torque sensor, and a monitoring screen. The motor prototype was connected to the torque sensor through the tooling, and a power tester was used to measure the motor power and current during the entire test process. All test data were displayed on the monitor screen in real time, as shown in Figure 7b. The torque sensor, power tester, and the software we used are all from SUGAWARA company, using non-inductive FOC to control them. During the experiment, the motor was kept running at a speed of 2200 rpm and the external load of the motor was gradually increased to investigate the changes in the current and output power of the motor from normal operation to hindered operation.

(a)      (b)

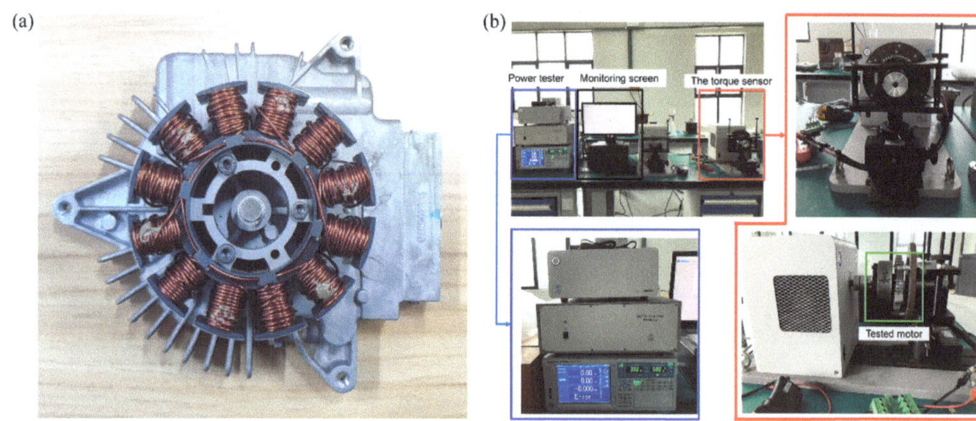


**Figure 7.** (**a**) Prototype, (**b**) schematic diagram of the bench test.

Figure 8a shows the corresponding input current-torque curve of the prototype. Taking the two points in the graph with the highest current as an example, the output torque is 1.8001 N·m when the experimental input current is 45.394 A, and 1.7996 N·m when the simulated current is 49.1331 A. The difference in torque is 0.0005 N·m when there is a difference in current torque of 7%. When the experimental input current is 42.1151 A, the output torque is 1.5443 N·m, and when the simulated current is 42.476 A, the output torque is 1.7098 N·m. When the current difference is 0.36 A, the torque difference is 10%, and the error is within the acceptable range. The reason for this error between experimental data and simulation data is mainly due to the fact that the motor adopts the external circuit as the excitation source when simulations are carried out, so that there is a certain gap between the resistance value of the armature winding in the simulation and that in the experiment, and the winding part in the external circuit cannot be fully equivalent to the motor winding. This leads to a certain difference between the experimental data and the simulation data. However, the experimental results verify the conclusion drawn from the simulation that the design of the motor can meet the requirements and can output a maximum torque of 1.8 N·m at 2200 rpm. At the same time, a temperature change diagram of each part of the motor was output by the temperature sensor, as shown in Figure 8b. The temperature of the motor gradually increases at the beginning of the experiment, and then tends to be stable. The temperature figure is based on the measured temperature of the motor under extreme operating conditions; the motor is not required to operate at this temperature for a long period of time and the motor can withstand this temperature due to its class F insulation material. Therefore, we consider that this temperature is acceptable. Figure 8c shows the efficiency of the motor based on the experiment data. From this image, it can be seen that the efficiency of the motor can reach 0.85; this efficiency is within the acceptable range due to the low cost of the motor. Therefore, the rationality of the simulation of the design through ANSYS Maxwell has been further corroborated.

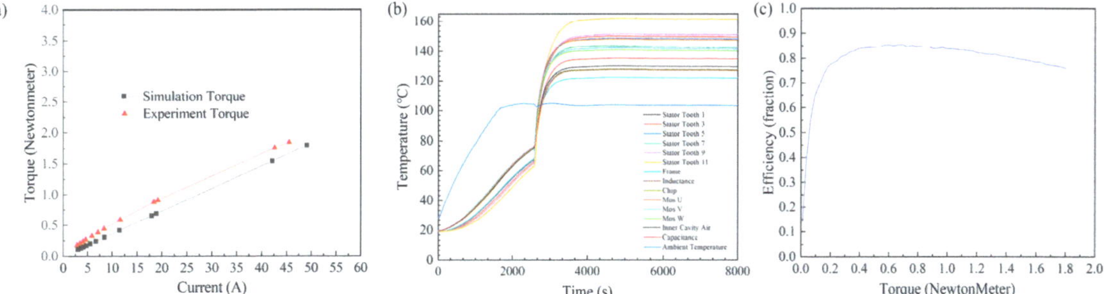

**Figure 8.** (**a**) Simulated and experimental current-torque curves, (**b**) temperature change diagram, and (**c**) efficiency waveform.

## 5. Conclusions

After the basic design of a PMBLDC motor, this paper simulates the load capability of the motor running at a fixed speed and input voltage by parameterizing the resistors in the control circuit. It is difficult to change the input current without changing the input voltage in brushless DC motor simulations. In this paper, by parameterizing the resistor in the control circuit of a PMBLDC motor, the input current can be adjusted without changing the input voltage; via this method, we have successfully simulated a motor load capacity test experiment in the simulation software, providing a new idea for subsequent simulation analyses of permanent magnet brushless DC motors.

**Author Contributions:** Conceptualization, K.R. and H.C.; methodology, K.R.; software, Q.W.; validation, H.S. and Q.S.; formal analysis, H.S.; investigation, K.R.; resources, K.R.; data curation, Q.W.; writing—original draft preparation, K.R. and H.C.; writing—review and editing, B.J.; visualization, B.J.; supervision, K.R.; project administration, Q.S.; funding acquisition, K.R. All authors have read and agreed to the published version of the manuscript.

**Funding:** The authors would like to thank the Natural Science Foundation of Jiangsu (No. BK20220407).

**Data Availability Statement:** Not applicable.

**Conflicts of Interest:** Author Bo Jin is employed by Huzhou Yueqiu Motor Co., Ltd. The remaining authors declare that the research was conducted in the absence of any commercial or financial relationships that could be construed as a potential conflict of interest.

## Abbreviations

The following abbreviations are used in this manuscript:

| | |
|---|---|
| DC | Direct current |
| NEV | New electric vehicles |
| BLDC | Brushless direct current |
| PMBLDC | Permanent magnet brushless direct current |
| FFT | Fast Fourier transform |
| RMS | Root mean square |
| FOC | Field oriented control |

## References

1. Gao, J.B.; Chen, H.B.; Tian, G.H.; Ma, C.C.; Zhu, F. An analysis of energy flow in a turbocharged diesel engine of a heavy truck and potentials of improving fuel economy and reducing exhaust emissions. *Energy Conv. Manag.* **2019**, *184*, 456–465. [CrossRef]
2. Guo, H.; Li, J.; Zhu, Z.; Lv, P. Numerical investigation on thermal performance of battery module with LCP-AFC cooling system applied in electric vehicle. *Energy Sci. Eng.* **2022**, *10*, 4431–4446. [CrossRef]
3. Li, Q.; Xu, Y.; Bu, S.; Yang, J. Smart vehicle path planning based on modified PRM algorithm. *Sensors* **2022**, *22*, 6581. [CrossRef]

4.  Song, S.; Zhou, H.; Jia, Z.; Xu, L.; Zhang, C.; Shi, M.; Hu, G. Effects of cutting parameters on the ultimate shear stress and specific cutting energy of sisal leaves. *Biosyst. Eng.* **2022**, *218*, 189–199. [CrossRef]
5.  Zheng, W.; Sun, J.; Ma, C.; Yu, Q. Percolation interpretation of film pressure forming mechanism of mechanical seal and calculation method of film pressure coefficient. *Tribol. Int.* **2022**, *173*, 107664. [CrossRef]
6.  Lu, S.M. A review of high-efficiency motors: Specification, policy, and technology. *Renew. Sust. Energ. Rev.* **2016**, *59*, 1–12. [CrossRef]
7.  Ntawiheba, J.D.; Qiao, L.; Niyonteze, J.D.D. Design and performance analysis of a high-efficiency Y3 induction motor. *J. Phys. Conf. Ser.* **2022**, *2345*, 012010. [CrossRef]
8.  Cipollone, R.; Di Battista, D. Sliding vane rotary pump in engine cooling system for automotive sector. *Appl. Therm. Eng.* **2015**, *76*, 157–166. [CrossRef]
9.  Zhu, G.; Liu, X.; Li, L.; Chen, H.; Tong, W.; Zhu, J. Cooling System Design of a High-Speed PMSM Based on a Coupled Fluidic-Thermal Model. *IEEE Trans. Appl. Supercond.* **2019**, *29*, 1–5. [CrossRef]
10. Tang, R.Y. *Modern Permanent Magnet Machines—Theory and Design*; China Machine Press: Beijing, China, 1997.
11. Zhu, G.J.; Jia, N.; Li, L.; Liu, T.; Xue, M.; Li, M.J. Cooling System Design Optimization of a High Power Density PM Traction Motor for Electric Vehicle Applications. *J. Electr. Eng. Technol.* **2021**, *16*, 3061–3068. [CrossRef]
12. Wang, W.; Yu, K.Y.; Huang, Q.L. Modeling and simulation of control strategy of engine multi-loop cooling system. *Ferroelectrics* **2021**, *579*, 33–43. [CrossRef]
13. Iora, P.; Tribioli, L. Effect of Ambient Temperature on Electric Vehicles' Energy Consumption and Range: Model Definition and Sensitivity Analysis Based on Nissan Leaf Data. *World Electr. Veh. J.* **2019**, *10*, 2. [CrossRef]
14. Lombardi, S.; Villani, M.; Chiappini, D.; Tribioli, L. Cooling System Energy Consumption Reduction through a Novel All-Electric Powertrain Traction Module and Control Optimization. *Energies* **2021**, *14*, 33. [CrossRef]
15. Wang, T.W.; Wagner, J. Advanced automotive thermal management—Nonlinear radiator fan matrix control. *Control Eng. Pract.* **2015**, *41*, 113–123. [CrossRef]
16. Feng, L.; Wikander, J.; Li, Z.W. Fuel Minimization of the Electric Engine Cooling System With Active Grille Shutter by Iterative Quadratic Programming. *IEEE Trans. Veh. Technol.* **2020**, *69*, 2621–2635. [CrossRef]
17. Lajunen, A.; Yang, Y.Y.; Emadi, A. Recent Developments in Thermal Management of Electrified Powertrains. *IEEE Trans. Veh. Technol.* **2018**, *67*, 11486–11499. [CrossRef]
18. Lelkes, A.; Bufe, M. IEEE BLDC motor for fan application with automatically optimized commutation angle. In Proceedings of the 35th Annual IEEE Power Electronics Specialists Conference (PESC 04), IEEE, Aachen, Germany, 20–25 June 2004; pp. 2277–2281.
19. Chakkarapani, K.; Thangavelu, T.; Dharmalingam, K.; Thandavarayan, P. Multiobjective design optimization and analysis of magnetic flux distribution for slotless permanent magnet brushless DC motor using evolutionary algorithms. *J. Magn. Magn. Mater.* **2019**, *476*, 524–537. [CrossRef]
20. Tang, Z.H.; Chen, Y.T.; Jiang, L.L.; Liang, R.H. Analysis and Design of a Smart Startup Method for a Single-Phase BLDC Fan Motor. *Electr. Power Compon. Syst.* **2018**, *46*, 1844–1856. [CrossRef]
21. Mayer, J.S.; Wasynczuk, O. Analysis and modeling of a single-phase brushless DC motor drive system. *IEEE Trans. Energy Convers.* **1989**, *4*, 473–479. [CrossRef]
22. Jeon, K.; Yoo, D.; Park, J.; Lee, K.D.; Lee, J.J.; Kim, C.W. Reliability-Based Robust Design Optimization for Maximizing the Output Torque of Brushless Direct Current (BLDC) Motors Considering Manufacturing Uncertainty. *Machines* **2022**, *10*, 797. [CrossRef]
23. Bharanigha, V.; Shuaib, Y.M. Minimization of torque ripples with optimized controller based four quadrant operation & control of BLDC motor. *Adv. Eng. Softw.* **2022**, *172*, 103192.
24. Krzysztof, S.; Anna, F.; Sławomir, W. Comparison of the Design of 3-Pole BLDC Actuators/Motors with a Rotor Based on a Single Permanent Magnet. *Sensors* **2022**, *22*, 3759.
25. Murat, T. Comparative analysis of the magnet effects on the permanent magnet BLDC motor performance used in electric vehicles. *Electr. Eng.* **2022**, *104*, 3411–3423.
26. Nejat, S.; Stefan, L.; Felix, K.; Annette, M. Noise and vibration characteristics of sub-fractional horsepower single-phase BLDC drives. *E & I Elektrotechnik und Informationstechnik* **2022**, *139*, 260–270.
27. Wang, X.H. *Permanent-Magnet Machine*; China Electric Power Press: Beijing, China, 2007.
28. Hendershort, J.R.; Miller, T.J.E. *Design of Brushless Permanent-Magnet Motors*; Magana Physics/Clarendon: Oxford, UK, 1994.
29. El-Refaie, A.M.; Shah, M.R.; Alexander, J.P.; Galioto, S.; Huh, K.K.; Gerstler, W.D. Rotor End Losses in Multi-Phase Fractional-Slot Concentrated-Winding Permanent Magnet Synchronous Machines. In Proceedings of the IEEE Energy Conversion Congress and Exposition (ECCE), Atlanta, GA, USA, 12–16 September 2010; IEEE: Piscataway, NJ, USA, 2010; pp. 1312–1320.
30. Shi, Z.; Sun, X.D.; Cai, Y.F.; Tian, X.; Chen, L. Design optimisation of an outer-rotor permanent magnet synchronous hub motor for a low-speed campus patrol EV. *IET Electr. Power Appl.* **2020**, *14*, 2111–2118. [CrossRef]

*World Electric Vehicle Journal*

*Article*

# Comparison and Analysis of Electromagnetic Characteristics of Basic Structure of Wireless Power Coil for Permanent Magnet Motors in Electric Vehicles

**Junfeng Zhao** [1,2,*], **Lingyun Zhao** [3], **Yuwei Zou** [4] **and Tianjin Chen** [5]

[1] Key Laboratory of Mold Intelligent Manufacturing of Taizhou, Taizhou 318020, China
[2] School of Mechatronics and Mould Engineering, Taizhou Vocational College of Science & Technology, Taizhou 318020, China
[3] School of Mechanical Engineering, Dalian University of Technology, Dalian 116023, China
[4] School of Physics and Electrical Engineering, Anyang Normal University, Anyang 455000, China
[5] Xuji Power Co., Ltd., Xuchang 461000, China
* Correspondence: zhao.zjf@gmail.com

**Abstract:** The purpose of this paper is to compare and analyze the electromagnetic characteristics of the basic structure of the coil in the electromagnetic coupling mechanism for the wireless power supply of permanent magnet motors in electric vehicles. The electromagnetic coupling mechanism is one of the key technologies for wireless power transmission and the coil structure plays a key role in the transmission performance of the coupling mechanism, and different structures can achieve different performances. The central objective of coil structure studies is to investigate how the coupling coefficient can be increased to achieve greater transmitted power and higher efficiency. In this paper, we investigate two basic coil configurations, circular and square, by studying their flux density variations when used as transmitting coils and their electromagnetic coupling characteristics when used as receiving coils. Three couplers consisting of circular and square coils are also analyzed in simulations and experiments are carried out on couplings containing circular and square coils of the same area. The results of the study show that the qualitative analysis, simulation analysis and experimental results are in high agreement. The results of this paper are an important reference for the design and optimization of wireless power coils for permanent magnet motors in electric vehicles.

**Keywords:** basic coil structure; electromagnetic characteristics; qualitative analysis; wireless power transfer

**Citation:** Zhao, J.; Zhao, L.; Zou, Y.; Chen, T. Comparison and Analysis of Electromagnetic Characteristics of Basic Structure of Wireless Power Coil for Permanent Magnet Motors in Electric Vehicles. *World Electr. Veh. J.* **2023**, *14*, 199. https://doi.org/10.3390/wevj14080199

Academic Editor: Ziqiang Zhu

Received: 6 June 2023
Revised: 17 July 2023
Accepted: 21 July 2023
Published: 26 July 2023



## 1. Introduction

With the popularization of electric vehicles and advancements in wireless power transfer technology, magnetic field-coupled static wireless charging technology has attracted significant attention in the field of electric vehicles. However, charging issues still remain one of the main bottlenecks in the development of electric vehicles. Therefore, researchers are actively exploring the possibility of wireless power transfer for electric vehicle permanent magnet motors [1–4].

As a hot topic in current research, there are still many unresolved issues regarding the coupling mechanism of wireless charging technology for electric vehicles. Some of these issues include the design of high-power coupling mechanisms suitable for fast charging modes, improving efficiency, reducing size and achieving lightweight construction.

The electromagnetic coupling mechanism is one of the key technologies for wireless power supply, and coil parameters are critical factors that affect the performance of the coupling mechanism. The coil structure plays a crucial role in the transmission performance, and different structures can achieve different performance characteristics [5–7]. The power transfer efficiency is not directly proportional to the coil size, and it needs to be analyzed on a case-by-case basis. Optimal coil parameters should be determined to achieve efficient power transmission while ensuring high-power transfer capability. In designing coils, studying how to improve the coupling coefficient is a crucial objective. A higher coupling coefficient implies higher transmission power and greater efficiency.

Researchers have already investigated circular pad couplers and found that the basic magnetic flux path within the pad remains relatively fixed [8]. Therefore, to enhance the coupling efficiency of coils, further research is needed into other coil structures and design methods. In 2010, the University of Auckland proposed a coil design called "Flux Pipe" based on their research on the magnetic field characteristics of circular coil pads, as shown in Figure 1. This design utilizes a dual-coil structure where two coils are connected in parallel, and the magnetic flux generated by the two coils is connected in series within the magnetic circuit. As a result, the design enhances the horizontal displacement tolerance [9]. In 2012, Saitama University in Japan introduced a transformer design called the "H-coil". This coil design is compact and tightly arranged, resulting in a smaller size, as shown in Figure 2. Additionally, the magnetic field of the coil is located on both sides of the coil, providing strong resistance against displacement [9,10].

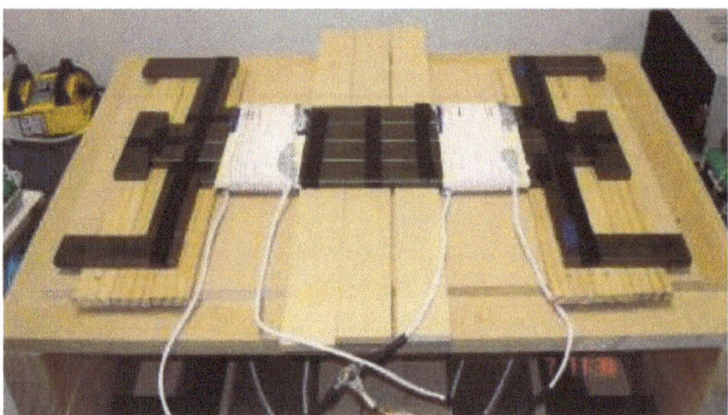

**Figure 1.** Prototype flux tube concept with coils in parallel.

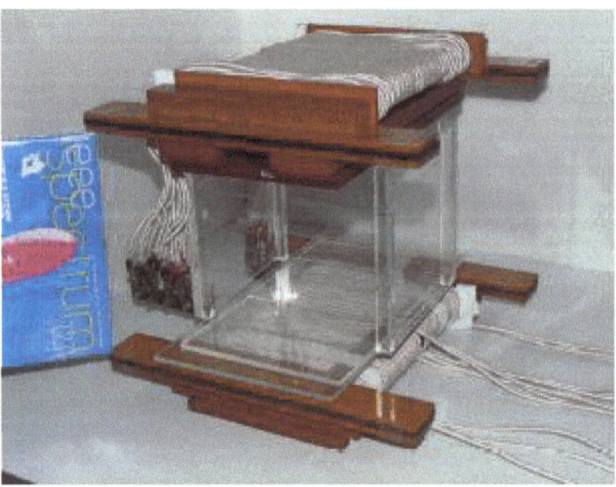

**Figure 2.** Photograph of the 3 kw transformer.

In 2013, the University of Auckland proposed a design scheme called the "DDQ" coil. This coil consists of two coils: a transmitting coil and a receiving coil. The transmitting coil adopts a DD-type coil, as shown in Figure 3, while the receiving coil utilizes a coil known as the DDQ coil, as shown in Figure 4. The distinguishing feature of this design is that the magnetic field directions are opposite, while the current directions at the center conductor are the same, resulting in enhanced magnetic field and improved displacement tolerance for electric vehicles. However, this design also faced the issue of magnetic field zero points. Subsequently, researchers introduced a modified design called the "DDD coil", which addressed the problem of magnetic field zero points. The DDD coil design effectively resolved the issue while reducing coil losses [7]. A. Zaheer proposed a design scheme for a bipolar BP coil, as shown in Figure 5. Compared to the DDQ coil, this design can reduce coil losses while achieving similar functionality [11]. In 2014, South Korea proposed a design scheme for a three-coil structure, as shown in Figure 6. This design consists of three coils, with one large coil nested inside a smaller coil. This design enhances the coil's resistance to interference and improves its stability [12].

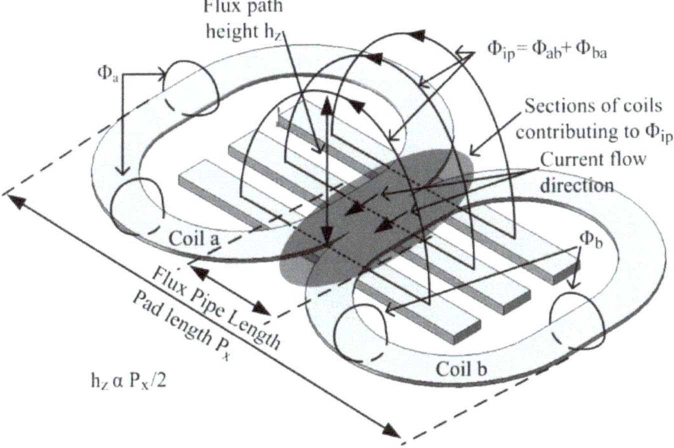


**Figure 3.** Simplified model of a DD pad.

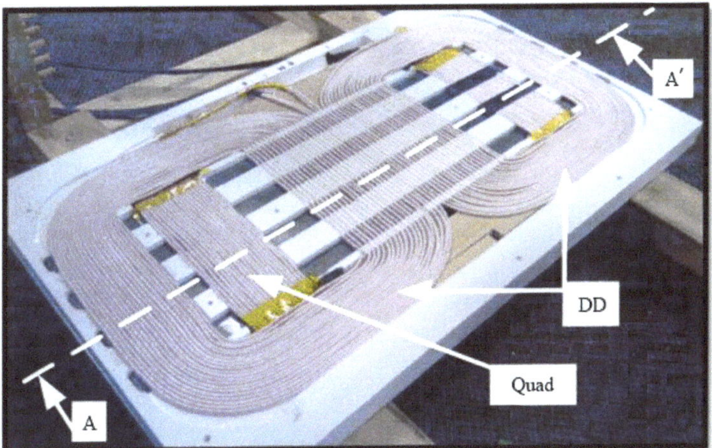


**Figure 4.** DD-Q pads with built receiver.

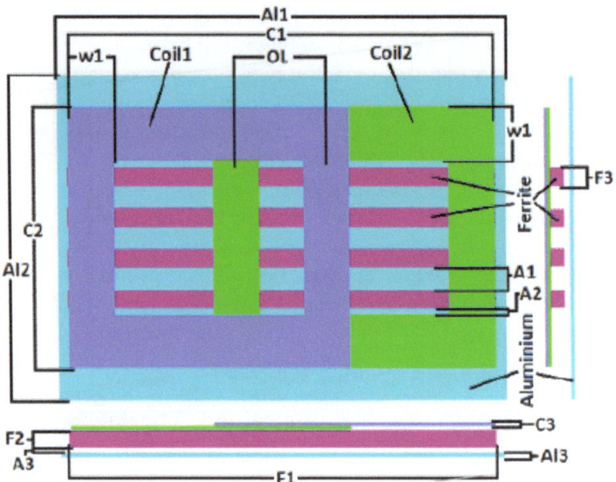



**Figure 5.** BPP construction.

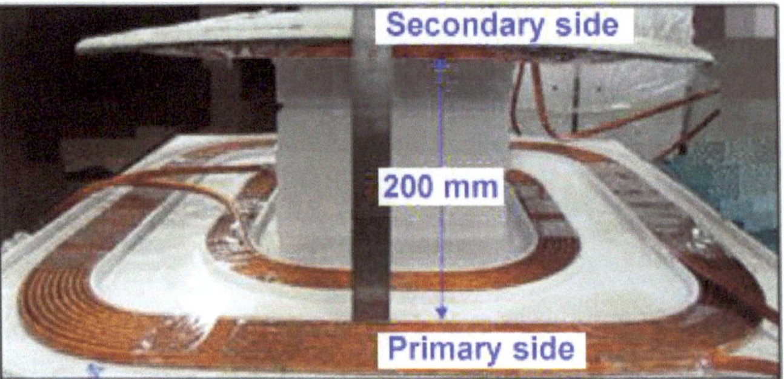



**Figure 6.** Three-coil structure.

In 2017, the University of Auckland proposed a design scheme called the TPP coil, as shown in Figure 7. Unlike traditional coil designs, the TPP coil utilizes a unique magnetic circuit structure that enhances the coil's resistance to displacement [13]. In 2019, the Tianjin University of Technology proposed a design called the "Taiji coil", based on the DD coil, as shown in Figure 8. The Taiji coil demonstrates similar transmission efficiency to circular coils and DD coils when in the centered position. However, in the misaligned state, it achieves a higher coupling coefficient, resulting in improved transmission efficiency. This demonstrates that the Taiji coil possesses better misalignment tolerance [14].

Nowadays, many types of couplers have been developed, such as square coils, circular pad, four square coils, DD couplers and DDQ couplers [4,5]. A comparison between circular and elliptical coils is made by studying the shape of the magnetic field and the path of the magnetic flux with a flux pipe concept [5,6]. The experimental results show that the coupling coefficient of the circular pad and DD coupling coils is higher and the cost is lower with respect to the coil type. Two analytical models of square and circular planar spiral coils were developed to compare their differences [10]. The results show that the square coil is a better choice for pad design when the outer diameter of the circular coil is equal to the outer side length of the square coil [15]. The characteristics of the coil structure are shown in Table 1.

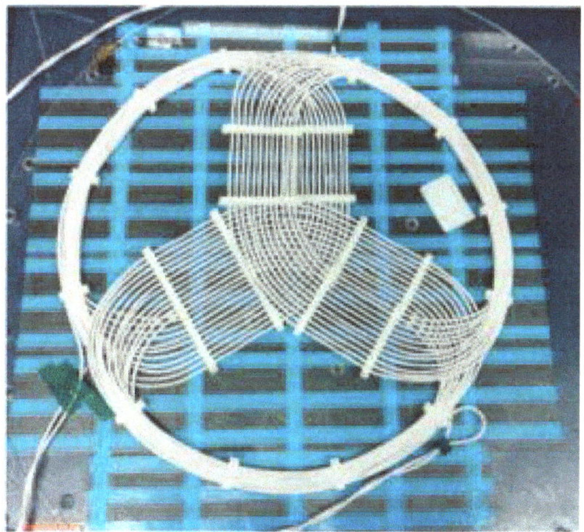

**Figure 7.** Prototype of TPP.

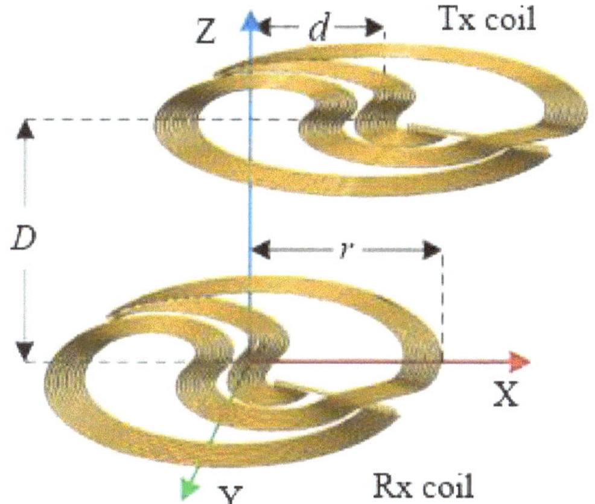


**Figure 8.** Taichi coil.

**Table 1.** The characteristics of coil structure.

| Coil Type | Offset Resistance | Magnetic Leakage | Losses (nH) | Coefficient (%) |
|---|---|---|---|---|
| Flux pipe | better | larger | larger | 42 (offset) |
| H | better | larger | larger | 90 |
| DD | better | smaller | average | 95.66 |
| DDQ | better | smaller | smaller | 95 |
| BP | average | smaller | smaller | 92.85 |

The above papers present specific coil designs and analyses based on the experience of the authors. There is no basic design guide. All kinds of more complex coils are composed of the two basic coil structures, circular and square. In this paper the problems of which is the better choice between circular and square structure and which one is more favorable to the system will be studied using basic circular and square coils. We encountered the above coil design problem when designing a wirelessly powered electromagnetic coupling mechanism for an electric bus. Our experimental subject was a Yutong ZK6875BEVG pure electric bus with a permanent magnet synchronous motor model YTM280-CV4-H.

For the two most common types of circular and square coils, the structure is axisymmetric. For each of the above two types of coils as the transmitting coil, the case of magnetic field coupling using two types of single-turn coils as the receiving coil is considered, as shown in Figure 9. First, the calculation models of the magnetic induction intensity of two kinds of coils were established, numerical calculations were carried out via MATLAB R2020b, and then the electromagnetic coupling characteristics of the two coils as receiving coils under three different conditions were qualitatively analyzed. Secondly, the simulation analysis of three kinds of couplers, a coupler with two circular loops (circular coupler), a coupler with one circular and one square loop (hybrid coupler), and a coupler with two square loops (square coupler), were carried out in Ansoft Maxwell 16.0 electromagnetic simulation software. Finally, the analysis was verified experimentally, and the characteristics of the two types of coils and the occasions for adjustment were obtained. The significance of this paper is that it establishes a reference for basic structural options that can be integrated into the selection and design of coil structures.

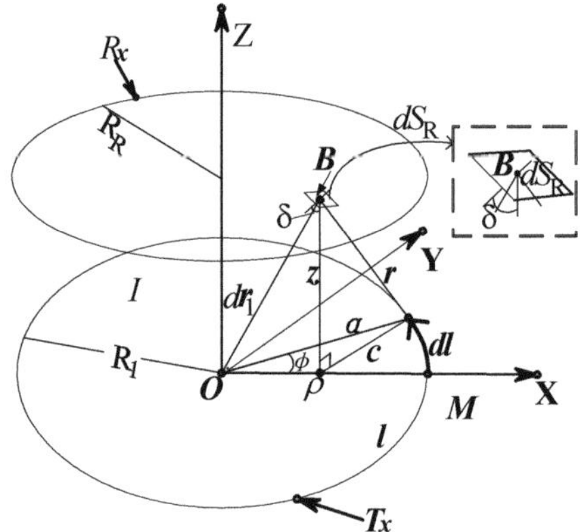

(**a**) A circular coupler.

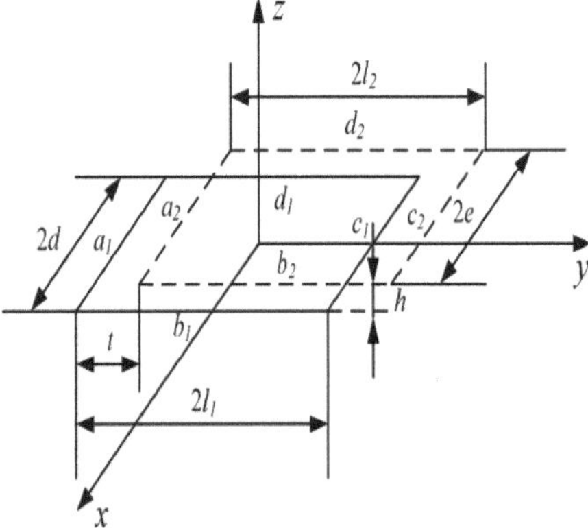

(**b**) A rectangular coupler.

**Figure 9.** Single-loop coupler.

## 2. Modeling and Analysis

For a given coil, according to Biot–Savart's law, the magnetic flux density of a line segment can be written as

$$\vec{B} = N_R N_T \frac{\mu_0 I(t)}{4\pi} \oint_l \frac{dl \times \vec{r}}{r} \tag{1}$$

where $N_R$ and $N_T$ are the turns of the receiving and transmitting coils, respectively, $I(t)$ is the current carried by the transmitting coil, $dl$ is the micro-element on the wire, and $r$ is the distance from the micro-element to one point in space.

### 2.1. Derivation of Circular Coil Magnetic Flux Density

The magnetic flux density $\vec{B}$ of the circular transmitting coil at any point in space is independent of the $\varphi$ coordinate due to the symmetry [16], as shown in Figure 9a.

$$\vec{B} = B_\rho \hat{\rho} + B_z \hat{z} \tag{2}$$

$$B_z|_{\rho=0} = \frac{\mu_0 I(t) a^2}{2(a^2 + z^2)^{3/2}} \tag{3}$$

$$B_z|_{\rho \neq 0} = \frac{\mu_0 I(t)}{2\pi\rho} \left( \frac{m}{4a\rho} \right)^{1/2} \left( \rho K(m) + \frac{am - (2-m)\rho}{2(1-m)} E(m) \right) \tag{4}$$

$$B_\rho = \frac{\mu_0 I(t) z}{2\pi\rho} \left( \frac{m}{4a\rho} \right)^{1/2} \left( \frac{2-m}{2(1-m)} E(m) - K(m) \right) \tag{5}$$

$$\begin{cases} K(m) = \int_0^{\pi/2} \frac{d\theta}{\sqrt{1 - m\sin^2\theta}} \\[2mm] E(m) = \int_0^{\pi/2} \sqrt{1 - m\sin^2\theta}\, d\theta \\[2mm] m = \frac{4a\rho}{z^2 + (a+\rho)^2} \end{cases} \tag{6}$$

where $B_r, B_z$ are the coordinate component of $B$. $a$, $z$, and $\rho$ are the length of the line in Figure 9a. In Figure 9a, $R_x$ is the receiving coil, $R_R$ is the radius of $R_x$, and the rest of the symbols are lengths or angles.

### 2.2. Derivation of Square Coil Magnetic Flux Density

The magnetic flux density at any point resulting from a line of finite length can be expressed as in [17], as shown in Figure 9b.

$$B_l = \frac{\mu_0 I}{4\pi\rho} \left( \frac{d+x}{\sqrt{\rho^2 + (d+x)^2}} + \frac{d-x}{\sqrt{\rho^2 + (d-x)^2}} \right) \tag{7}$$

where $d$ is the side length of square coil and $\rho$ is the distance from one point to the line segment. For a square coil, the magnetic flux density $\vec{B}$ can be expressed as

$$\vec{B} = \vec{B}_a + \vec{B}_b + \vec{B}_c + \vec{B}_d \tag{8}$$

where the subscripts a, b, c, and d represent the segments of the corresponding square coil.

### 2.3. Qualitative Analysis of Electromagnetic Properties

For the transmitting coils with a given surrounding area of 0.28 m$^2$, according to (2) and (8), the normalized curve of the magnetic flux density can be obtained when the circular and square coils are used as the transmitting coils, respectively, as shown in Figure 10. In Figure 10, the horizontal axis is the horizontal distance from any point of constant height to the central axis of the transmitter coil. The height of the measuring point is 0.25 m from the coils. It can be seen from Figure 10 that the B value is higher inside the coils and gradually decreases outside the coils.

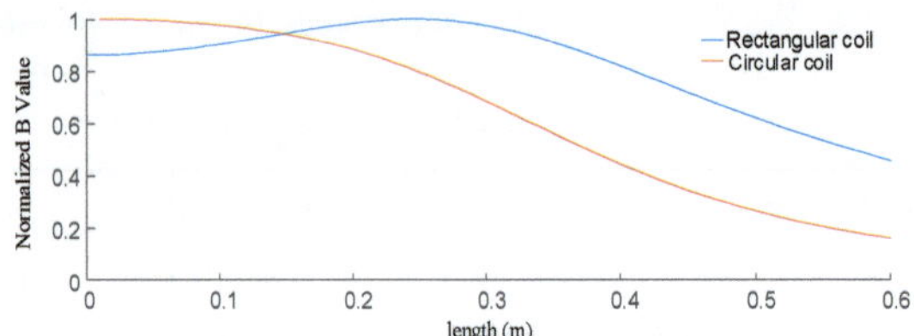

**Figure 10.** Normalized magnetic flux density curve.

For the circular and square loops serving as receiving coils, there are three scenarios: (1) the enclosed area of the two receiving coils is the same; (2) the circumference of the two receiving coils is the same; (3) the outer diameter of the two receiving coils is the same, as depicted in Figure 11a, b, and c, respectively.

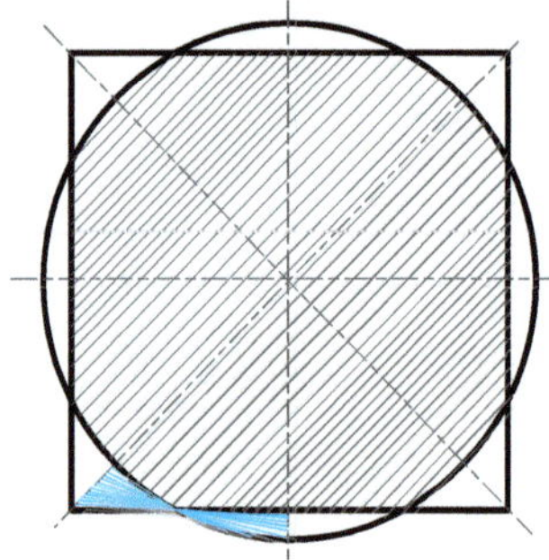

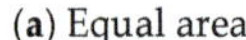

(**a**) Equal area.

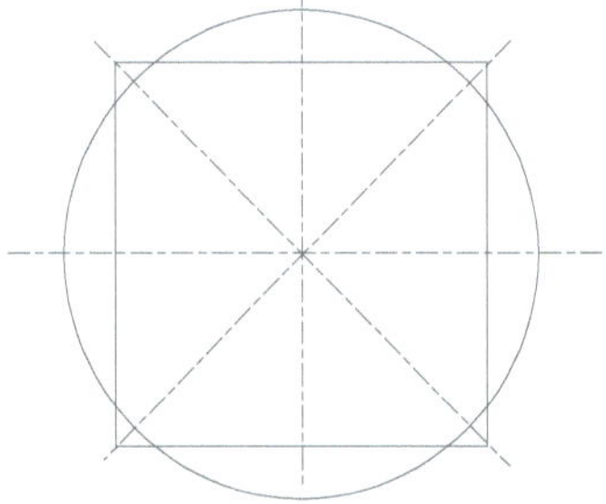

(**b**) Equal circumference.

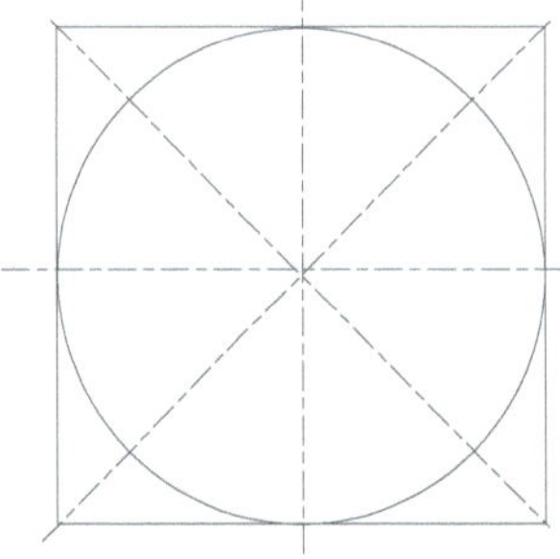

(**c**) Equal outer diameter.

**Figure 11.** Comparison of three cases.

In the case of an equal surrounding area, the girth ratio is $\sqrt{\pi}/2$, as shown in Figure 11a. The lightly shaded area is the overlap of the two receiving coils. It is obvious that the magnetic flux of the two receiving coils is the same in this part. Due to symmetry, the dark shaded area of the two receiving coils is the same. According to the conclusion of the electromagnetic induction characteristics of the transmitting coil in Figure 10, the magnetic flux of the circular receiving coil is greater than or equal to that of the square receiving coil in the dark shaded area. Therefore, for circular and square receiving coils with the same surrounding area, the magnetic flux of the circular receiving coil should be greater than or equal to that of the square receiving coil.

For the case of the same perimeter as shown in Figure 11b, the area ratio is $4/\pi$. According to the above analysis result, the magnetic flux of the circular receiving coil is greater than that of the square receiving coil because the surrounded area of the circular receiving coil is larger than that of the square receiving coil.

In the case of the same outer diameter, as shown in Figure 11c, all of the area and circumference ratio is $\pi/4$. The magnetic flux of the circular receiving coil is smaller than that of the square receiving coil because the surrounded area of the circular receiving coil is smaller than that of the square receiving coil.

## 3. Numerical Simulations

The results of the above analysis are compared by simulating the mutual inductance of the coils using electromagnetic simulation software.

### 3.1. Normalized Coupling Coefficient Comparison

The results of the qualitative analysis show that the equal surrounded area result is in the middle of the other two results, so it is a key point of comparison for circular and square coils. Clear results can be obtained by analyzing circular and square coils of equal enveloping area. Figure 12 shows the simulation of normalized coupling coefficients for the same enveloping area of circular, hybrid, and square couplers of the three coupler types. The horizontal axis is the horizontal distance between the center of the transmitting coil and the center of the receiving coil. The size of the square loop is 0.3 m $\times$ 0.3 m in length and width, and the radius of the circular loop is 0.17 m. The height of the two coupling coils in the simulation is fixed at 0.3 m. The simulation results show that the coupling coefficient is highest for the circular coupler and lowest for the square coupler, i.e., the coupling coefficient of the circular coupler is higher than that of the square coupler. These results are consistent with the qualitative analysis above.

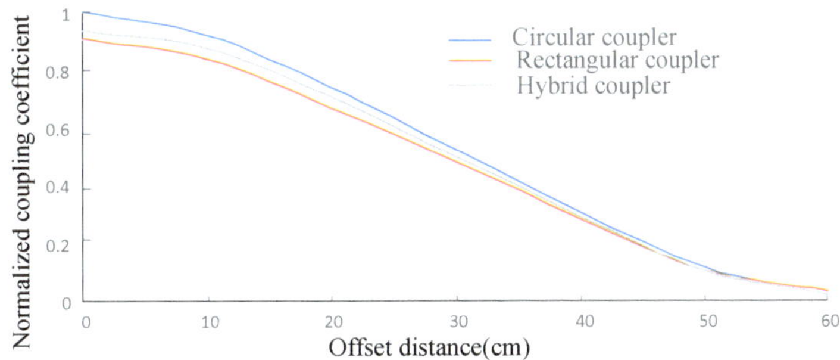


**Figure 12.** Normalized coupling coefficient curve.

### 3.2. Coupling Mutual Inductance Comparison

Table 2 shows the mutual inductances in the surrounding area of 0.09 m$^2$, 0.16 m$^2$, and 0.25 m$^2$ of circular coupler, hybrid coupler, and square coupler, respectively. It can clearly be seen that the mutual inductance and coupling coefficient of the coupler with the circular loop are larger than that with the square loop.

World Electr. Veh. J. **2023**, 14, 199

**Table 2.** Comparison of different area couplers s.

| Coil Type | Pacing (m) | Area (m²) | Mutual Inductance (nH) | Coupling Coefficient (%) |
|---|---|---|---|---|
| circular | | | 30 | 2.52 |
| hybrid | | 0.09 | 29 | 2.4 |
| rectangular | | | 28 | 2.23 |
| circular | | | 67.5 | 4.13 |
| hybrid | 0.3 | 0.16 | 66.9 | 3.94 |
| rectangular | | | 65.5 | 3.75 |
| circular | | | 120 | 5.62 |
| hybrid | | 0.25 | 119.19 | 5.4 |
| rectangular | | | 117.5 | 5.15 |

In Table 3, a square loop with a length of 0.3 m is used as a reference, and the circular and square loops are, respectively, used as the transmitting coil and the receiving coil to simulate the nine kinds of structure of the coupler with the same surrounded area, the same circumference, and the same outer diameter, respectively.

**Table 3.** Detailed simulation data comparison of couplers.

| Coil Type | Pacing (m) | Area (m²) | Perimeter (m) | Outer Diameter (m) | Mutual Inductance (nH) | Coupling Coefficient (%) |
|---|---|---|---|---|---|---|
| circular | | | 1.06 | 0.34 | 30 | 2.52 |
| hybrid | | 0.09 | | | 29 | 2.4 |
| rectangular | | | 1.2 | 0.3 | 28 | 2.23 |
| circular | | 0.11 | | 0.38 | 42.22 | 3.12 |
| hybrid | 0.3 | | 1.2 | | 34.95 | 2.69 |
| rectangular | | 0.09 | | 0.3 | 28 | 2.23 |
| circular | | 0.07 | 0.94 | | 19.99 | 1.97 |
| hybrid | | | | 0.3 | 23.7 | 2.1 |
| rectangular | | 0.09 | 1.2 | | 28 | 2.23 |

It can be seen from Table 2 that the mutual inductance of the circular coupler is greater than that of a square coupler. The concept of the self-inductance coefficient is employed to compare the difference between the coils of two couplers. The self-inductance of the circular coil, the mutual inductance, and the coupling coefficient of the circular coupler are set as $L_1$, $M_1$, and $K_1$, respectively. The self-inductance of the square coil, the mutual inductance, and the coupling coefficient of the square coupler are set as $L_2$, $M_2$, and $K_2$, respectively. The mutual inductance and the coupling coefficient of the hybrid coupler are set to $M_3$ and $K_3$, respectively.

The proportional relationship of the self-inductance of the coupled structure is obtained.

$$\frac{\sqrt{L_i L_j}}{\sqrt{L_k L_m}} = \frac{M_s K_t}{M_t K_s} \tag{9}$$

where $i, j, k, m \in (1,2)$, $s = \begin{cases} i, & i = j \\ 3, & i \neq j \end{cases}$, $t = \begin{cases} k, & k = m \\ 3, & k \neq m \end{cases}$.

For a set of coupling structures of equal surrounded area in Table 2, the self-inductance coefficient of a circular coupler and square coupler can be obtained from (9):

$$\frac{L_1}{L_2} = \frac{M_1 K_2}{M_2 K_1} \approx 0.95 \tag{10}$$

For a set of coupling structures of the same perimeter in Table 2, the self-inductance coefficient of a circular coupler and square coupler can be obtained from (9):

$$\frac{L_1}{L_2} = \frac{M_1 K_2}{M_2 K_1} \approx 1.08 \tag{11}$$

According to the data in Table 2 and the contrast between (10) and (11), it can be obtained that self-inductance is not only affected by the line length but also closely related to the structure. For the same length, the self-inductance of the circular coil inductance is larger than that of the square one.

## 4. Experimental Validation

In order to verify the theoretical and simulation results, we set up an experimental platform for three kinds of coupler.

Based on the above discussion, it can be seen that the results obtained under the condition that the square coil and the circular coil surround the same area are the watersheds of the three cases. If the results of equal area are obtained under the same conditions, the other two results are clear.

In this experiment, the outer diameter, wire diameter, and number of turns of the circular coil are 15.6 cm, 2.5 mm, and five turns, respectively. The self-inductance of the two circular coils is 7.49 uH and 7.42 uH, respectively. The outer diameter, wire diameter, and number of turns of the square coil are 13.8 cm, 2.5 mm, and five turns, respectively. The self-inductance of the two regulation coils is 7.49 uH and 7.42 uH, respectively.

The experimental platform is shown in Figure 13. The mutual inductance calculation method can be performed by measuring two coils in series and subtracting two self-inductances [13]. The experimental data and results are shown in Table 4. Table 3 lists the mutual inductance and coupling coefficients of the three types of couplers at three different pitches for the transmitter and receiver coils. Figure 14 shows the verification system for this experiment. Figure 15 shows the wireless power supply system and coil that we developed for the electric vehicle.

**Table 4.** Experimental data on equal surrounded areas.

| Coil Type | Pacing (mm) | Mutual Inductance (uH) | Coupling Coefficient (%) |
|---|---|---|---|
| circular | | 0.857 | 11.5 |
| hybrid | 67.82 | 0.844 | 11.21 |
| rectangular | | 0.735 | 9.66 |
| circular | | 0.609 | 8.17 |
| hybrid | 79.17 | 0.607 | 8.06 |
| rectangular | | 0.593 | 7.8 |
| circular | | 0.332 | 4.46 |
| hybrid | 105.17 | 0.329 | 4.37 |
| rectangular | | 0.326 | 4.29 |

(**a**) Equal area.

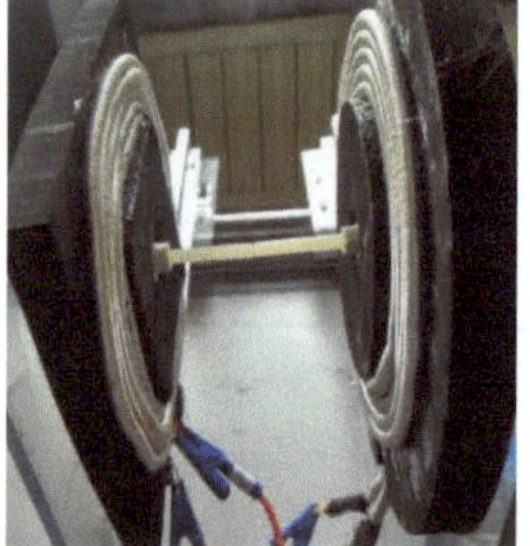

(**b**) Equal circumference.

(**c**) Equal outer diameter.

**Figure 13.** Comparison of three cases of experimental coils.

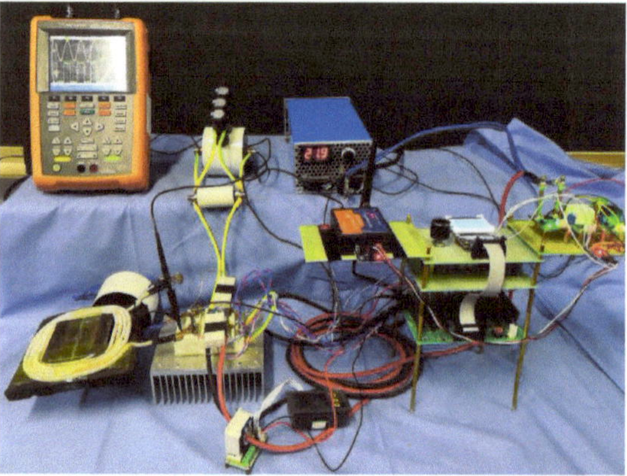

**Figure 14.** The experimental system.

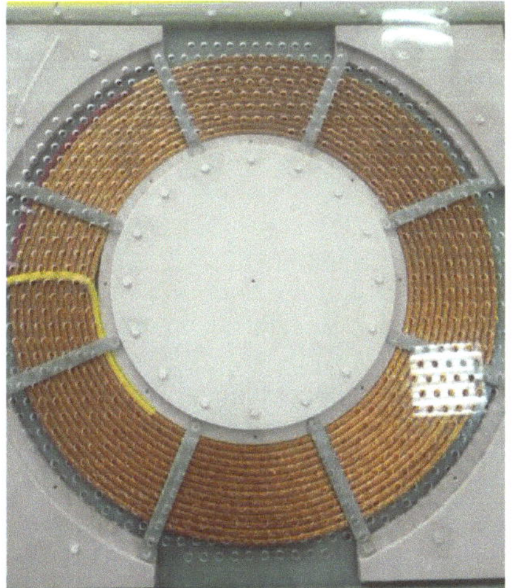

(**a**) Designed wireless power transfer coils for EV.

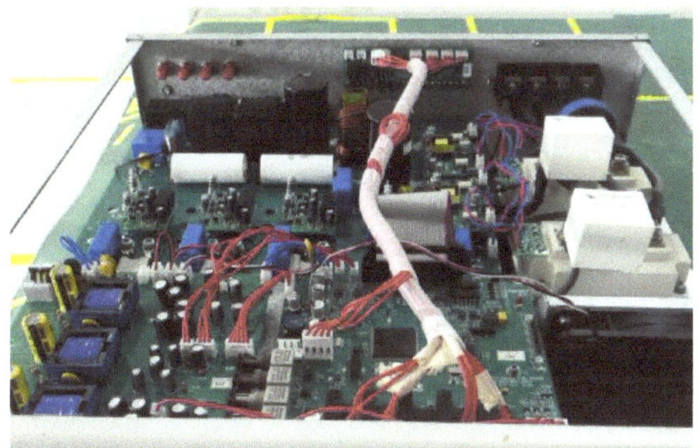

(**b**) Design of a wireless power supply transmitter for EV.

**Figure 15.** The wireless power supply system for EV.

## 5. Discussion of the Electromagnetic Characteristics of Two Basic Coil Shapes

According to the characteristic curve of the magnetic induction intensity of the transmitting coils, the electromagnetic coupling characteristics of the circular and square loops as the receiving coil under three different conditions were qualitatively analyzed by comparing the magnetic fluxes of the two coils for equal surrounded area, equal circumference, and equal outer diameter, respectively. The qualitative analysis result for equal surrounded area is in the middle of the other two results, so it is a key point of comparison for circular and square loops. Therefore, the simulation analysis is mainly based on the same surrounded area conditions of circular and square loops. Secondly, three kinds of coupler are simulated with the coil of equal surrounded area. The simulation results show that the mutual inductance coupling coefficient of the coupler with the circular loop is larger than that with the square coil. Finally, the mutual inductances of the three kinds of coupler for different surrounded area are compared, and the mutual inductances of the three kinds of coupler for the same surrounded area, the same circumference, and the same outer diameter are simulated, respectively. The final results show that the qualitative analysis and simulation analysis are highly consistent.

## 6. Conclusions

Through the above analysis, the circular coil flux is greater than the square coil flux in the case of equal surrounded area or equal perimeter. Namely, in these two cases, the circular receiving coil mutual inductance is greater than that of the square receiving coil. The cost and loss of the circular and square coil are equal when the circumference is equal. The mutual inductance of the circular coil is clearly better than that of the square coil, so the efficiency is higher. When the enclosed area is equal, the cost and loss of the circular coil is lower than that of the square coil, and the mutual inductance is slightly larger than that of the square coil, so the efficiency is also higher than that of the square coil. Therefore, it is best to choose a circular coil with no space constraints. Under the condition of limited space, the square coil can achieve greater mutual inductance, but the cost and loss of the coil will be higher than that of the circular coil. The above conclusions provide the basis for the development of more reasonable and efficient composite couplers by using the basic structure.

**Author Contributions:** Conceptualization, J.Z. and T.C.; methodology, Y.Z. and J.Z.; software, L.Z. and J.Z.; validation, Y.Z.; writing—original draft preparation, J.Z.; writing—review and editing, L.Z. All authors have read and agreed to the published version of the manuscript.

**Funding:** This work was supported by the National power grid headquarter science and technology project (funding number 5201011600TW).

**Data Availability Statement:** Some or all data and models generated or used during the study are available in a repository or online.

**Conflicts of Interest:** The authors declare no conflict of interest. Tianjin Chen is an employee of Xuji Power Co., Ltd. The paper reflects the views of the scientists and not the company.

## References


1. Zhao, Z.; Liu, F.; Chen, K. New progress of wireless charging technology for electric vehicles. *Trans. China Electrotech. Soc.* **2016**, *31*, 30–40. (In Chinese)
2. Assawaworrarit, S.; Yu, X.; Fan, S. Robust wireless power transfer using a nonlinear parity-time-symmetric circuit. *Nature* **2017**, *546*, 387. [CrossRef] [PubMed]
3. Kurs, A.; Karalis, A.; Moffatt, R.; Joannopoulos, J.D.; Fisher, P.; Soljacic, M. Wireless power transfer via strongly coupled magnetic resonances. *Science* **2007**, *317*, 83–86. [CrossRef] [PubMed]
4. Zhao, J.F.; Huang, X.L.; Wang, W. Efficiency analysis of magnetic resonance wireless power transfer with three-dimensional transmitters. *J. Appl. Phys.* **2015**, *117*, 83. [CrossRef]
5. Wake, K.; Laakso, I.; Hirata, A.; Chakarothai, J.; Onishi, T.; Watanabe, S.; De Santis, V.; Feliziani, M. Derivation of Coupling Factors for Different Wireless Power Transfer Systems: Inter- and Intralaboratory Comparison. *IEEE Trans. Electromagn. Compat.* **2017**, *59*, 677–685. [CrossRef]
6. Wang, Y.; Yao, Y.; Liu, X.; Xu, D. S/CLC Compensation Topology Analysis and Circular Coil Design for Wireless Power Transfer. *IEEE Trans. Transp. Electrif.* **2017**, *3*, 496–507. [CrossRef]
7. Budhia, M.; Boys, J.T.; Covic, G.A.; Huang, C.-Y. Development of a Single-Sided Flux Magnetic Coupler for Electric Vehicle IPT Charging Systems. *IEEE Trans. Ind. Electron.* **2013**, *60*, 318–328. [CrossRef]
8. Budhia, M.; Covic, G.A.; Boys, J.T. Design and optimization of circular magnetic structures for lumped inductive power transfer systems. *IEEE Trans. Power Electron.* **2011**, *26*, 3096–3108. [CrossRef]
9. Budhia, M.; Covic, G.; Boys, J. A new IPT magnetic coupler for electric vehicle charging systems. In Proceedings of the IECON 2010—36th Annual Conference on IEEE Industrial Electronics Society, Glendale, AZ, USA, 7–10 November 2010; pp. 2487–2492.
10. Takanashi, H.; Sato, Y.; Kaneko, Y.; Abe, S.; Yasuda, T. A large air gap 3 kW wireless power transfer system for electric vehicles. In Proceedings of the Energy Conversion Congress and Exposition, Raleigh, NC, USA, 15–20 September 2012; pp. 269–274.
11. Zaheer, A.; Kacprzak, D.; Covic, G.A. A bipolar receiver pad in a lumped IPT system for electric vehicle charging applications. In Proceedings of the 2012 IEEE Energy Conversion Congress and Exposition (ECCE), Raleigh, NC, USA, 15–20 September 2012; pp. 283–290.
12. Moon, S.; Kim, B.-C.; Cho, S.-Y.; Ahn, C.-H.; Moon, G.-W. Analysis and Design of a Wireless Power Transfer System with an Intermediate Coil for High Efficiency. *IEEE Trans. Ind. Electron.* **2014**, *61*, 5861–5870. [CrossRef]
13. Kim, S.; Covic, G.A.; Boys, J.T. Tripolar Pad for Inductive Power Transfer Systems for EV Charging. *IEEE Trans. Power Electron.* **2017**, *32*, 5045–5057. [CrossRef]
14. Li, Y.; Zhao, J.; Yang, Q.; Liu, L.; Ma, J.; Zhang, X. A Novel Coil with High Misalignment Tolerance for Wireless Power Transfer. *IEEE Trans. Magn.* **2019**, *55*, 2800904. [CrossRef]

15. Luo, Z.; Wei, X. Analysis of Square and Circular Planar Spiral Coils in Wireless Power Transfer System for Electric Vehicles. *IEEE Trans. Ind. Electron.* **2018**, *65*, 331–341. [CrossRef]
16. Zhao, J.; Huang, X.; Wang, W. Wireless power transfer with two-dimensional resonators. *IEEE Trans. Magn.* **2014**, *50*, 1–4. [CrossRef]
17. Zhao, J. Research on Three-Dimensional Wireless Power Transmission System. Ph.D. Thesis, Southeast University, Nanjing, China, 2015. (In Chinese)

*World Electric Vehicle Journal*

MDPI

*Article*

# Research on the dq-Axis Current Reaction Time of an Interior Permanent Magnet Synchronous Motor for Electric Vehicle

**Anxue Huang [1], Zhongxian Chen [1,*] and Juanjuan Wang [2]**

[1]  School of Intelligence Manufacturing, Huanghuai University, Zhumadian 463000, China
[2]  School of Computer and Artificial Intelligence, Huanghuai University, Zhumadian 463000, China
*  Correspondence: chenzhongxian@huanghuai.edu.cn

**Abstract:** An interior permanent magnet synchronous motor (IPMSM) is a kind of drive motor with high power density that is suitable for electric vehicles. In this paper, the dq-axis current reaction time of IPMSM was investigated in order to improve the reaction time of the electric vehicle. Firstly, the mathematical model of the current-loop decoupling of IPMSM was presented. Secondly, the controller design of dq-axis current-loop decoupling of IPMSM was investigated by the methods of proportional integral (PI) and internal model control PI (IMC-PI). Thirdly, based on the methods of PI and IMC-PI, the influence of the inverter switching frequency on the dq-axis current reaction time of IPMSM was analyzed and simulated, and it was found that the inverter switching frequency only had a significant influence on the parameters set of the PI controller. Lastly, compared with the PI method, the results of the simulation and hardware experiment demonstrate that the dq-axis current reaction time of IPMSM was improved by the IMC-PI method, and the IMC-PI method had the advantage of simple parameters setting and was not influenced by the inverter switching frequency.

**Keywords:** start-up response; synchronous motor; current-loop decoupling; electric vehicle; dq-axis current

**Citation:** Huang, A.; Chen, Z.; Wang, J. Research on the dq-Axis Current Reaction Time of an Interior Permanent Magnet Synchronous Motor for Electric Vehicle. *World Electr. Veh. J.* **2023**, *14*, 196. https://doi.org/10.3390/wevj14070196

Academic Editors: Zhongze Wu and Ziqiang Zhu

Received: 15 June 2023
Revised: 14 July 2023
Accepted: 20 July 2023
Published: 23 July 2023

## 1. Introduction

IPMSM is a rotor permanent magnet motor with permanent magnets embedded in the rotor [1–3]. With the advantages of high power density and high torque/inertia ratio, IPMSM is widely used in high-performance drive systems such as industrial robots, equipment manufacture, and electric vehicles. In order to improve the driving performance of IPMSM, the current-loop control of IPMSM were investigated in some literatures. The current-loop control of IPMSM is an output torque control strategy, which aims to improve the dynamic response capability of IPMSM, including the reaction time of the starting operation state and constant speed performance of the stable operation state [4,5].

Usually, the theoretical modeling and hardware experiment of current loop control of IPMSM are based on the dq synchronous rotating coordinate system [6,7]. In this coordinate system, the amplitudes of coupling voltages of IPMSM are determined by the torque (dq-axis current) and speed of IPMSM [8,9]. The reason is that the high torque or high speed will increase the amplitude of coupling voltages, which will seriously affect the performance of current-loop control of IPMSM. For the traditional controller design of the current loop of the motor, the common practice is to ignore the coupling voltages of dq synchronous rotating coordinate system [10–12]. In this condition, if the q-axis current changes, an error will occur in the d-axis current, which will lead to the distortion of the motor's output torque and dynamic response performance.

With the certain electrical parameters, the dq-axis coupling voltages of IPMSM can be eliminated by voltage feedforward decoupling control (VFDC). However, the electrical parameters of IPMSM depend on its operating conditions and control modes [13,14]. Therefore, the constant values setting of electrical parameters of IPMSM is not beneficial for the voltage decoupling of IPMSM thoroughly.

Based on the Popov hyperstability theory, Qiu T. et al. proposed an adaptive observer to monitor the permanent magnet flux linkage of PMSM, and the experimental results showed that the sensitivity of q-axis inductance of PMSM was decreased by the adaptive observer and adaptation proportional–integral (PI) controller [15,16]. However, with the increase in the PMSM speed, the q-axis inductance errors of the adaptive observer increased. Saleh M. and Hassan M. et al. proposed a comprehensive control method including the fusion of the sliding-mode method and type-2 neuro-fuzzy systems to control the speed of the induction motor (or the doubly fed induction generator). The analysis and comparison results indicated that the adaptive sliding-mode type-2 neuro-fuzzy controller can control the induction motor (or the doubly fed induction generator) with higher performance (compared with type-1 neuro-fuzzy systems) [17,18]. However, the speed fluctuation of the induction motor still existed, and the reaction time of the motor was not analyzed in detail. Xu W. et al. proposed a novel sliding-mode-based extended state observer (SMESO) to improve the dynamic response capability of the permanent magnet synchronous motor (PMSM). After the signal was input into the feed-forward compensation controller, the comprehensive simulation and experiment's results show that the dynamic response capability of PMSM can be improved [19]. Furthermore, some improved methods based on the field-oriented control (FOC), direct torque control (DTC), and sliding-mode observer (SMO) were proposed to investigate the dynamic response of PMSM [20–24]. Nevertheless, it is necessary to explore a simple method to improve the reaction time of PMSM on the basis of traditional PI control technology.

In order to improve the dq-axis current reaction time of IPMSM with a simplified voltage-decoupling controller, PI and IMC-PI methods were used to investigate the current-loop control of IPMSM in this paper. The main contributions of this paper are as follows. Firstly, the theoretical model of current-loop decoupling of IPMSM was analyzed and presented. Secondly, the controller design of current-loop decoupling of IPMSM was investigated, including the methods of PI and IMC-PI. Thirdly, based on the methods of PI and IMC-PI, the influence of inverter switching frequency on the dq-axis current reaction time of IPMSM was simulated and analyzed. The analysis results showed that the inverter switching frequency had a significant influence on the parameters set of the PI controller. Lastly, the results of the simulation calculation and hardware experiment demonstrated that compared with the PI method, the dq-axis current reaction time of IPMSM was improved by the IMC-PI method, and the IMC-PI method had the advantage of the easy setting of the control parameters.

## 2. Mathematical Model of the Current Loop of IPMSM

In the dq synchronous rotating coordinate system, if the magnetic saturation, eddy current loss, and copper loss are ignored, then the voltage equation of *d*-axis and *q*-axis of IPMSM can be described as

$$
\begin{cases}
u_d = Ri_d + L_d \frac{d}{dt} i_d - \omega_e \psi_q \\
u_q = Ri_q + L_q \frac{d}{dt} i_q + \omega_e \psi_d
\end{cases}
\tag{1}
$$

where the subscript symbols $d$ and $q$ are the d-axis and q-axis, respectively, $u_d$ and $u_q$ are the voltages, $i_d$ and $i_q$ are the currents, $L_d$ and $L_q$ are the inductances, $\psi_d$ and $\psi_q$ are the flux linkages, $R$ is the resistance of stator phase winding, and $\omega_e$ is the electric angular velocity of rotor of IPMSM [25–27]. In addition, the d-axis flux linkage $\psi_d$ and q-axis flux linkage $\psi_q$ can be written as

$$
\begin{cases}
\psi_d = L_d i_d + \psi_f \\
\psi_q = L_q i_q
\end{cases}
\tag{2}
$$

where $\psi_f$ is the excitation flux linkage of permanent magnets of IPMSM.

Substitute Equation (2) into Equation (1), then the voltage equation of IPMSM can be rewritten as

$$\begin{cases} u_d = Ri_d + L_d \frac{d}{dt} i_d - \omega_e L_q i_q \\ u_q = Ri_q + L_q \frac{d}{dt} i_q + \omega_e \left( L_d i_d + \psi_f \right) \end{cases} \tag{3}$$

Equation (3) indicates that the q-axis coupling term $\omega_e L_q i_q$ exists in the d-axis voltage $u_d$, and the d-axis coupling term $\omega_e L_d i_d$ exists in the q-axis voltage $u_q$, and thus the couple voltages occurred in the dq synchronous rotating coordinate system of IPMSM. Furthermore, the amplitude of dynamic term $\omega_e \psi_f$ is decided by the electric angular velocity $\omega_e$ (namely the speed of rotor of IPMSM).

In order to design the controller of the current loop of IPMSM, the decoupling of coupling terms $\omega_e L_q i_q$ and $\omega_e L_d i_d$ should be completed. Therefore, by adopting the decoupling method that ignores the coupling terms $\omega_e L_q i_q$ and $\omega_e L_d i_d$, and also ignores the dynamic term $\omega_e \psi_f$, Equation (3) can be simplified as

$$\begin{cases} u_d = Ri_d + L_d \frac{d}{dt} i_d \\ u_q = Ri_q + L_q \frac{d}{dt} i_q \end{cases} \tag{4}$$

After the Laplace transform [28], the transfer function between IPMSM's dq-axis current and dq-axis voltage can be described as

$$\begin{cases} G_d(s) = \frac{i_d(s)}{u_d(s)} = \frac{1}{L_d s + R} \\ G_q(s) = \frac{i_q(s)}{u_q(s)} = \frac{1}{L_q s + R} \end{cases} \tag{5}$$

Based on Equation (5) and automatic control theory, the traditional PI controller of the current loop of IPMSM can be obtained as

$$\begin{cases} u_d^* = \left( k_{pd} + \frac{k_{id}}{s} \right) \left( i_{dref} - i_d \right) - \omega_e L_q i_q \\ u_q^* = \left( k_{pq} + \frac{k_{iq}}{s} \right) \left( i_{qref} - i_q \right) + \omega_e \left( L_d i_d + \psi_f \right) \end{cases} \tag{6}$$

where $k_{pd}$ and $k_{id}$ are the proportional coefficient and integral coefficient of the d-axis current loop, and $i_{dref}$ and $i_d$ are the set value and feedback value of the d-axis current. In the same way, $k_{pq}$ and $k_{iq}$ are the proportional coefficient and integral coefficient of the q-axis current loop, and $i_{qref}$ and $i_q$ are the set value and feedback value of q-axis current.

## 3. Controller Design of Current-Loop Decoupling of IPMSM

### 3.1. PI Controller Design

Considering the transfer function of d-axis voltage $G_d(s)$ and inverter $K_{pwm} / \left( T_{pwm} s + 1 \right)$ [29], the d-axis PI current-loop structure of IPMSM is shown in Figure 1a. In this paper, the transfer function of d-axis voltage $G_d(s)$ is derived from the mathematical model of the current loop of IPMSM (see Section 2 of this paper), and the transfer function of inverter $K_{pwm} / \left( T_{pwm} s + 1 \right)$ can be assumed as a pure lag amplification link with delay. Additionally, the transfer function of q-axis voltage $G_q(s)$ is same with the transfer function of d-axis voltage $G_d(s)$ (see Equation (5)), and thus the q-axis PI current-loop structure of IPMSM is the same as Figure 1a, as shown in Figure 1b.

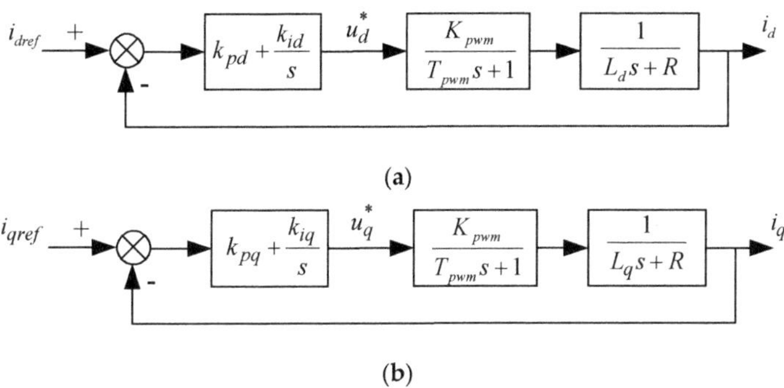

**(a)**

**(b)**

**Figure 1.** The PI current-loop structure of IPMSM: (**a**) d-axis PI current-loop structure; (**b**) q-axis PI current-loop structure.

In Figure 1, $K_{pwm}$ and $T_{pwm}$ are the magnification factor and switching cycle of the inverter, respectively. Taking the d-axis PI current-loop structure of IPMSM as an example, the loop transfer function of Figure 1a can be written as

$$G_d(s) = \frac{k_{pd}(\tau_i s + 1)}{\tau_i s} \frac{K_{pwm}}{T_{pwm}s + 1} \frac{1/R}{\frac{L_d}{R}s + 1}\bigg|_{\tau_i = k_{pd}/k_{id}} \tag{7}$$

It is assumed that $\tau_i = k_{pd}/k_{id} = L_d/R$, where $\tau_i = k_{pd}/k_{id}$ is the structure transformation of the PI controller and $\tau_i = L_d/R$ is the structure transformation of the transfer function of IPMSM; then, the extreme points of the d-axis transfer function $G_d(s)$ of Equation (7) can be eliminated. In this condition, Equation (7) can be regarded as the typical I type system, as shown in Equation (8).

$$G_d(s) = \frac{k_{pd}K_{pwm}}{L_d(T_{pwm}s + 1)s} = \frac{K_I}{(T_{pwm}s + 1)s} \tag{8}$$

In Equation (8), the parameter $K_I$ is described as $K_I = k_{pd}K_{pwm}/L_d$. According to the motion control theory of typical I type system [29,30], if the desired overshoot of the d-axis current of IPMSM is less than 5%, then the parameter $K_I$ should be selected as

$$K_I = \frac{1}{2T_{pwm}} \tag{9}$$

Under situation of $K_I = k_{pd}K_{pwm}/L_d = 1/(2T_{pwm})$, the proportional coefficient $k_{pd}$ and integral coefficient $k_{id}$ of the d-axis current loop of IPMSM can be calculated as

$$k_{pd} = \frac{L_d}{2T_{pwm}K_{pwm}} \tag{10}$$

$$K_{id} = \frac{K_{pd}}{\tau_i} = \frac{R}{2T_{pwm}K_{pwm}} \tag{11}$$

In the same way, the proportional coefficient $k_{pq}$ and integral coefficient $k_{iq}$ of the q-axis current loop of IPMSM can be inferred as

$$K_{pq} = \frac{L_q}{2T_{pwm}K_{pwm}} \tag{12}$$

$$K_{iq} = \frac{K_{pq}}{\tau_i} = \frac{R}{2T_{pwm}K_{pwm}} \tag{13}$$

In addition, the above PI parameters ($k_{pd}$, $k_{iq}$, $k_{pq}$, and $k_{iq}$) of the dq-axis current loop of IPMSM also can be inferred and obtained by the classical automatic control theory. Taking the d-axis PI current-loop structure of IPMSM as an example, based on Equation (8), the d-axis closed-loop transfer function of Figure 1a can be written as

$$\Phi_d(s) = \frac{k_{pd}k_{pwm}}{L_dT_{pwm}s^2 + L_ds + k_{pd}K_{pwm}} = \frac{K_g}{s^2 + \frac{1}{T_{pwm}}s + K_g} \bigg|_{K_g = \frac{k_{pd}K_{pwm}}{L_dT_{pwm}}} \tag{14}$$

In the classic control theory of some references [31,32], the transfer function of the second-order system can be written as

$$\Phi(s) = \frac{\omega_n^2}{s^2 + 2\xi\omega_n s + \omega_n^2} \tag{15}$$

Therefore, the relationship between the transfer function of the second-order system and typical type I system can be obtained by the comparison of Equations (14) and (15)

$$\omega_n = \sqrt{K_g} = \frac{1}{2\xi T_{pwm}} \tag{16}$$

After combining Equations (14) and (16), the proportional coefficient $k_{pd}$ and integral coefficient $k_{id}$ of Equation (14) can be obtained as

$$K_{pd} = \frac{L_d}{4\xi^2 T_{pwm}K_{pwm}} \tag{17}$$

$$K_{id} = \frac{K_{pd}}{\tau_i} = K_{pd}\frac{R}{L_d} = \frac{R}{4\xi^2 T_{pwm}K_{pwm}} \tag{18}$$

If the damping coefficient is $\xi = 0.707$ (ideal value), then the proportional coefficient $k_{pd}$ and integral coefficient $k_{id}$ of the d-axis current loop of IPMSM can be further refined as

$$\begin{cases} K_{pd} = \frac{L_d}{2T_{pwm}K_{pwm}} \\ K_{id} = \frac{R}{2T_{pwm}K_{pwm}} \end{cases} \tag{19}$$

Similarly, the proportional coefficient $k_{pq}$ and integral coefficient $k_{iq}$ of the q-axis current loop of IPMSM can be inferred as

$$\begin{cases} K_{pq} = \frac{L_q}{2T_{pwm}K_{pwm}} \\ K_{iq} = \frac{R}{2T_{pwm}K_{pwm}} \end{cases} \tag{20}$$

The comparison of Equations (10)–(13), (19) and (20) indicates that the above two PI parameter tuning results of IPMSM are the same. Actually, the above two PI parameter tuning results of IPMSM all originated from the dynamic performance indicators and empirical formulas of the typical type I system.

### 3.2. IMC-PI Controller Design

In Section 3.1, the PI parameter tuning result of IPMSM is under the condition of full dq-axis current-loop decoupling. Actually, the existence of salient pole characteristics in IPMSM determines that the dq-axis current loop cannot be fully decoupled. This phenomenon of non-fully decoupled dq-axis current loop is not beneficial to the high-precision and low-sensitivity parameters setting of the controller. In order to ensure the high precision and low sensitivity of the parameters setting of the controller, the IMC method was adopted in this paper to improve traditional PI controller design of the current loop of IPMSM, which is named as the IMC-PI method. Furthermore, it should be noted that the aim of the low-sensitivity parameters setting of the controller is to improve the operational stability (anti-interference ability) of IPMSM. Therefore, there is no contradiction between the high precision and low sensitivity of the parameters setting of the controller.

Figure 2 shows the structure of IMC, where the parameters of Figure 2 are described in Table 1.

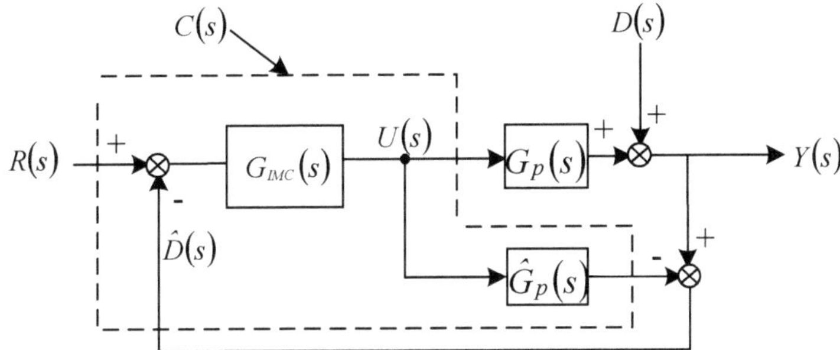


**Figure 2.** The structure of IMC.

**Table 1.** The parameters of IMC.

| Parameters | Descriptions |
| --- | --- |
| $R(s)$ | System's input signal |
| $C(s)$ | Feedback controller |
| $\hat{D}(s)$ | Feedback signal |
| $G_{IMC}(s)$ | IMC controller |
| $G_p(s)$ | System's model |
| $\hat{G}_p(s)$ | Internal model |
| $D(s)$ | Interference signal |
| $Y(s)$ | System's output signal |

In Figure 2, the node signal $U(s)$ can be described as

$$U(s) = \frac{(R(s) - D(s))G_{IMC}(s)}{1 + (G_p(s) - \hat{G}_p(s))G_{IMC}(s)} \tag{21}$$

and the system's output signal $Y(s)$ is

$$Y(s) = G_p(s)U(s) + D(s) \tag{22}$$

According to Equations (21) and (22), the relationship among the system's output signal $Y(s)$, system's input signal $R(s)$, and interference signal $D(s)$ can be expressed as

$$Y(s) = \frac{G_p(s)G_{IMC}(s)}{1 + (G_p(s) - \hat{G}_p(s))G_{IMC}(s)}R(s) + \frac{1 - G_{IMC}(s)\hat{G}_p(s)}{1 + (G_p(s) - \hat{G}_p(s))G_{IMC}(s)}D(s) \tag{23}$$

Based on Equation (23), the structure of IMC (see Figure 2) can be simplified, as shown in Figure 3. In Figure 3, the feedback controller $C(s)$ can be described as

$$C(s) = \frac{G_{IMC}(s)}{1 - G_{IMC}(s)\hat{G}_p(s)} \tag{24}$$

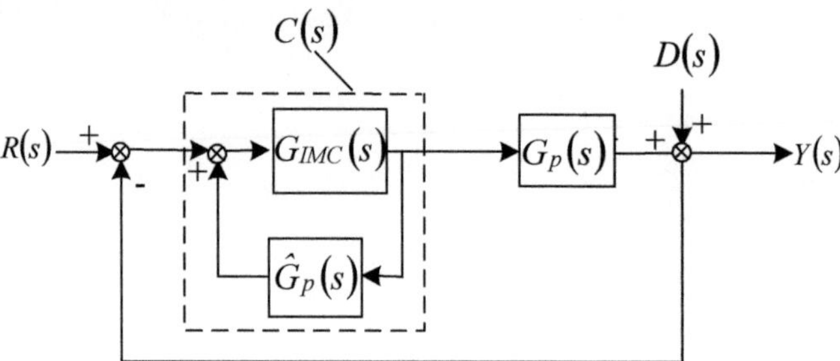

**Figure 3.** Equivalent structure simplification of IMC.

If there is no interference signal $D(s)$, and $G_{IMC}(s) = \hat{G}_p(s)^{-1}$, then the relationship between the system's input signal $R(s)$ and system's output signal $Y(s)$ can be expressed as

$$Y(s) = G_{IMC}(s)G_p(s)R(s) = G_p(s)^{-1}G_p(s)R(s) = R(s) \tag{25}$$

From Equation (25), it can be concluded that the IMC method is beneficial to ensure the consistence between the system's input signal $R(s)$ and system's output signal $Y(s)$.

Regardless of the interference signal $D(s)$, the d-axis IMC-PI current-loop structure of IPMSM is shown in Figure 4.

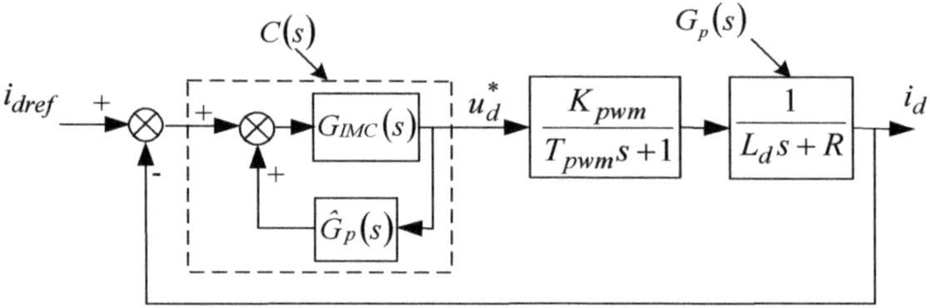

**Figure 4.** The *d*-axis IMC-PI current-loop structure of IPMSM.

It is assumed that $G_p(s) = \hat{G}_p(s)$, and

$$G_{IMC}(s) = \hat{G}_p^{-1}(s)L(s) = G_p^{-1}(s)L(s) \tag{26}$$

where $L(s) = \varepsilon/(s + \varepsilon)$, and $\varepsilon$ is the modulation parameter of the d-axis IMC-PI current-loop structure of IPMSM. On this basis, substitute Equation (26) into Equation (24), then the feedback controller $C(s)$ and the parameter tuning results of the d-axis IMC-PI current-loop structure of IPMSM can be obtained as

$$C(s) = \varepsilon\left(L_d + \frac{R}{s}\right) \tag{27}$$

$$\begin{cases} K_{pd(IMC)} = \varepsilon L_d \\ K_{id(IMC)} = \varepsilon R \end{cases} \tag{28}$$

The comparison results between Equations (19) and (28) indicates that the number of parameter tuning is decreased from two ($T_{pwm}$ and $K_{pwm}$ of PI current-loop structure) to one ($\varepsilon$ of IMC-PI current-loop structure), and thus the difficulty of parameter tuning is also decreased by the IMC-PI current-loop structure of IPMSM.

Similarly, the parameter tuning results of the q-axis IMC-PI current-loop structure of IPMSM can be inferred as

$$\begin{cases} K_{pq(IMC)} = \varepsilon L_q \\ K_{iq(IMC)} = \varepsilon R \end{cases} \tag{29}$$

From Equations (28) and (29), it can be concluded that responsiveness of the dq-axis current of IPMSM can be improved by the larger modulation parameter $\varepsilon$ of the IMC-PI current-loop structure. However, the larger modulation parameter $\varepsilon$ will also increase the overshoot and stability time of the dq-axis current. Therefore, the selection of modulation parameter $\varepsilon$ of the IMC-PI current-loop structure should be solved reasonably.

### 3.3. The Modulation Parameter Selection of IMC-PI

In the practical engineering application, the current loop of IPMSM can be transformed and approximated to one order system. If the inverter transfer function $K_{pwm} / (T_{pwm}s + 1)$ is ignored, then the d-axis open-loop transfer function $G_{d1}(s)$ and d-axis closed-loop transfer function $\Phi_{d1}(s)$ of IPMSM can be described as

$$G_{d1}(s) = \left( K_{pd} + \frac{K_{id}}{s} \right) \frac{1}{L_d s + R} \tag{30}$$

$$\Phi_{d1}(s) = \frac{K_{pd}s + K_{id}}{L_d s^2 + K_{pd}s + Rs + K_{id}} \tag{31}$$

Substitute the proportional coefficient $K_{pd(IMC)}$ and integral coefficient $K_{id(IMC)}$ of Equation (28) into Equation (31), then the Equation (31) can be simplified as

$$\Phi_{d1}(s) = \frac{\varepsilon}{s + \varepsilon} = \frac{1}{Ts + 1} = \left. \frac{1}{\frac{1}{\varepsilon}s + 1} \right|_{T = \frac{1}{\varepsilon}} \tag{32}$$

where Equation (32) is a closed-loop transfer function with one order system, and its open-loop system is a typical I type system.

According to the definition of bandwidth frequency $\omega_b$ of typical Type I systems

$$20lg|\Phi(j\omega_b)| = 20lg\frac{1}{\sqrt{1 + T^2\omega_b^2}} = 20lg\frac{1}{\sqrt{2}} \Rightarrow \omega_b = \frac{1}{T} = \varepsilon \tag{33}$$

the modulation parameter $\varepsilon$ of the IMC-PI current-loop structure is also the bandwidth frequency of the closed-loop transfer function of IPMSM. Moreover, the transfer function $1/(L_d s + R)$ of the d-axis current loop of IPMSM can be regarded as a $RL$ system, which is composed of d-axis inductance $L_d$ and resistance $R$. If it is assumed that the parameter $T = L_d / R$, then the modulation parameter $\varepsilon$ of the IMC-PI current-loop structure can be further described as

$$\varepsilon = 2\pi \frac{R}{L_d} \tag{34}$$

where the coefficient $2\pi$ converts the unit of modulation parameter $\varepsilon$ of the IMC-PI current-loop structure from frequency $H_z$ to radians per second rad/s.

Considering that IPMSM has both the d-axis current loop and q-axis current loop, the modulation parameter $\varepsilon$ of the IMC-PI controller is calculated as

$$\varepsilon = 2\pi\min\left( \frac{R}{L_d}, \frac{R}{L_q} \right) \tag{35}$$

## 4. Simulation Analysis and Experimental Verification

### 4.1. Simulation Analysis

In order to analyze the difference between the PI method and IMC-PI method, and verify the advantage of the IMC-PI method in improving the dq-axis current reaction time of IPMSM, some simulation analysis is carried out in this section. The electrical parameters of IPMSM are shown in Table 2, and the dq-axis current-loop structure of IPMSM is shown in Figure 5.

**Table 2.** Electrical parameters of IPMSM.

| Item | Value | Unit |
|---|---|---|
| Number of phases | 3 | - |
| Power | 1.5 | kW |
| Rated line voltage | 220 | V |
| Rated line current | 4.5 | A |
| Rated speed | 2000 | r/min |
| Number of stator slots | 18 | - |
| Number of rotor poles | 8 | - |
| Air-gap length | 0.5 | mm |
| Phase resistance | 2.92 | $\Omega$ |
| D-axis inductance | 8.96 | mH |
| Q-axis inductance | 12.29 | mH |
| Excitation flux linkage | 0.955 | Wb |

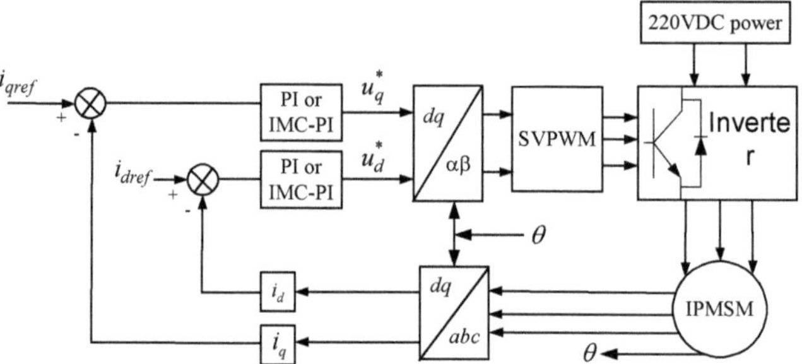


**Figure 5.** The dq-axis current-loop structure of IPMSM.

Based on the Equations (19), (20), (28) and (29), and the electrical parameters of IPMSM (see Table 2), the parameter tuning results of the PI method and IMC-PI method are shown in Table 3, and the relevant simulation analysis results of the dq-axis current reaction time of IPMSM are shown in Figures 6–8. In the process of simulation analysis, the magnification factor of the inverter is $K_{pwm} = 1$, and the IPMSM's speed is zero (namely, the IPMSM's rotor is in the locked condition).

**Table 3.** Parameter tuning results of the PI method and IMC-PI method.

| | Item | $T_{pwm} = 0.001s$ | $T_{pwm} = 0.01s$ | $T_{pwm} = 0.1s$ |
|---|---|---|---|---|
| | $K_{pq}$ | 6.145 | 0.6145 | 0.06145 |
| PI method | $K_{iq}$ | 1460 | 146 | 14.6 |
| | $K_{pd}$ | 4.48 | 0.448 | 0.0448 |
| | $K_{id}$ | 1460 | 146 | 14.6 |
| | $K_{pq(IMC)}$ | 18.3 | 18.3 | 18.3 |
| IMC-PI method | $K_{iq(IMC)}$ | 4356 | 4356 | 4356 |
| | $K_{pd(IMC)}$ | 13.3 | 13.3 | 13.3 |
| | $K_{id(IMC)}$ | 4356 | 4356 | 4356 |

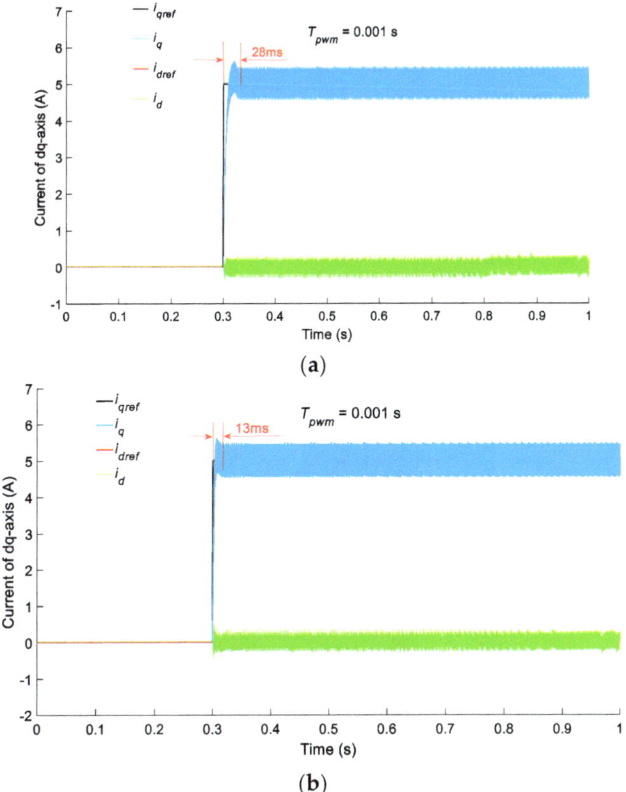

**Figure 6.** The simulation analysis results of the dq-axis current reaction time of IPMSM ($T_{pwm} = 0.001s$): (**a**) PI method; (**b**) IMC-PI method.

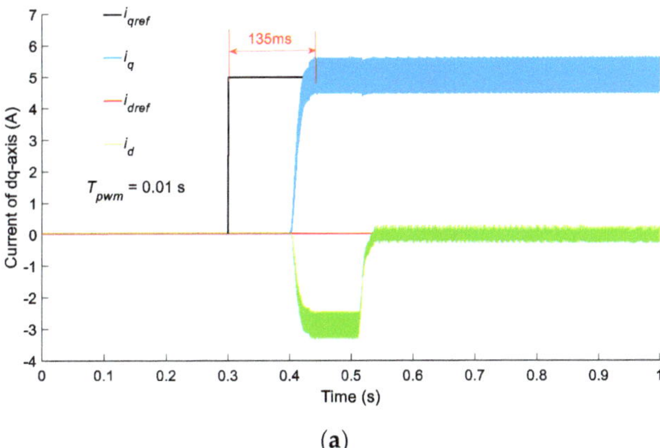

**Figure 7.** *Cont.*

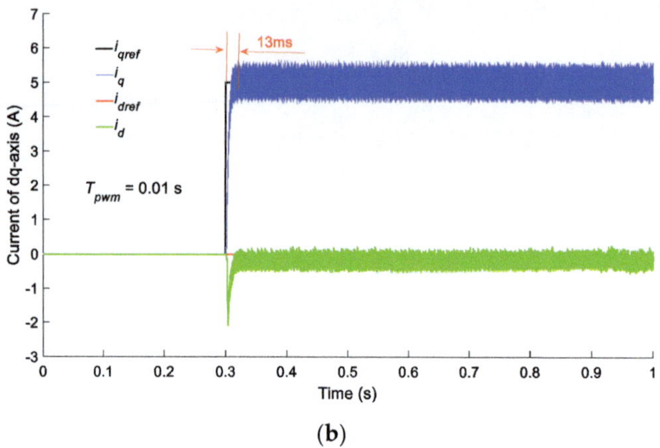

**(b)**

**Figure 7.** The simulation analysis results of the dq-axis current reaction time of IPMSM ($T_{pwm} = 0.01s$): (**a**) PI method; (**b**) IMC-PI method.

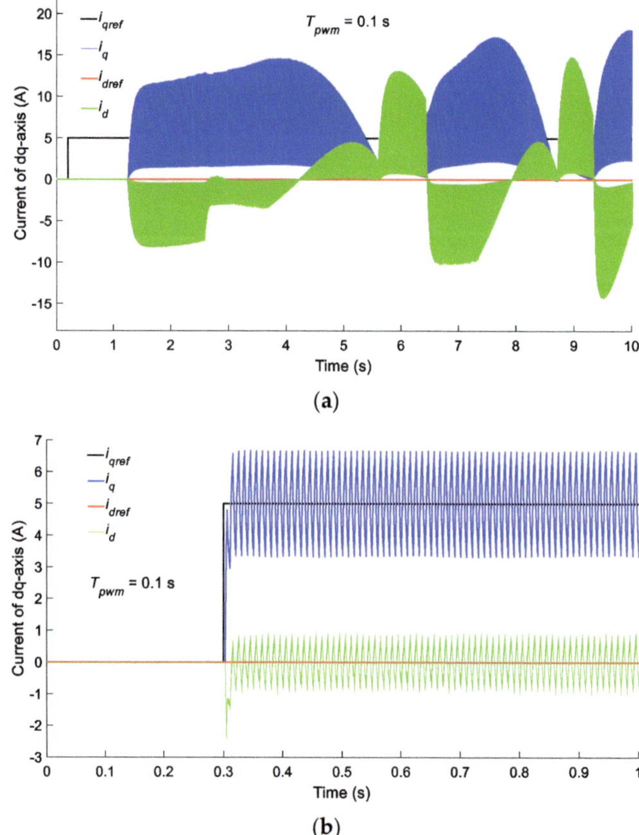

**(a)**

**(b)**

**Figure 8.** The simulation analysis results of the dq-axis current reaction time of IPMSM ($T_{pwm} = 0.1s$): (**a**) PI method; (**b**) IMC-PI method.

In Figures 6–8, the step pulse amplitude of the q-axis target current is $i_{qref} = 5A$, and the d-axis target current is $i_{dref} = 0A$. From Figures 6–8, it can be concluded that with the IMC-PI method, the switching cycle $T_{pwm}$ of the inverter almost has no impact on the dq-axis current reaction time of IPMSM ($T_{pwm} = 0.001s$, $T_{pwm} = 0.01s$, and $T_{pwm} = 0.1s$, respectively). However, when the switching cycle $T_{pwm}$ of the inverter varies from 0.001 s to 0.1 s, the dq-axis current reaction time of IPMSM is decreased by the PI method. For example, based on the PI method and switching cycle $T_{pwm} = 0.01s$, the delay time of the dq-axis current reaction time of IPMSM is approximately 0.1 s, as shown in Figure 7a. What is more important is that when the switching cycle is $T_{pwm} = 0.1s$, the delay time of the dq-axis current reaction time of IPMSM is increased to 1 s (see Figure 8a).

In addition, as the switching cycle $T_{pwm}$ increases, the stability of the dq-axis current of IPMSM decreases. For example, the mutual influence between the q-axis feedback current $i_q$ and d-axis feedback current $i_d$ occurs (see Figures 7a and 8a), and even the distortion of the q-axis feedback current $i_q$ and d-axis feedback current $i_d$ occurs (see Figure 8a).

The comparison results among Figures 6–8 indicate that by the PI method, the switching cycle $T_{pwm}$ of the inverter plays an important role in the dq-axis current reaction time of IPMSM. However, the IMC-PI method can ensure a fine dq-axis current reaction time of IPMSM with a different switching cycle $T_{pwm}$ of the inverter. These conclusions are consistent with the above analysis results (the comparison among Equations (19), (20), (28) and (29)).

### 4.2. Hardware Experimentation

Figure 9 shows the IPMSM's structure and test platform, where the electrical parameters of IPMSM are the same as Table 2. In Figure 9b, the rotor position of IPMSM is measured by the rotary encoder (2500 lines of resolution ratio), where the rotary encoder is coaxial with the rotor of IPMSM and installed in the back-end department of the rotor. The 2500 lines mean that if the rotor of IPMSM rotates one revolution (360 degrees), the rotary encoder will generate 2500 pulse signals. The more lines of rotary encoder, the more precise the position of the rotor.

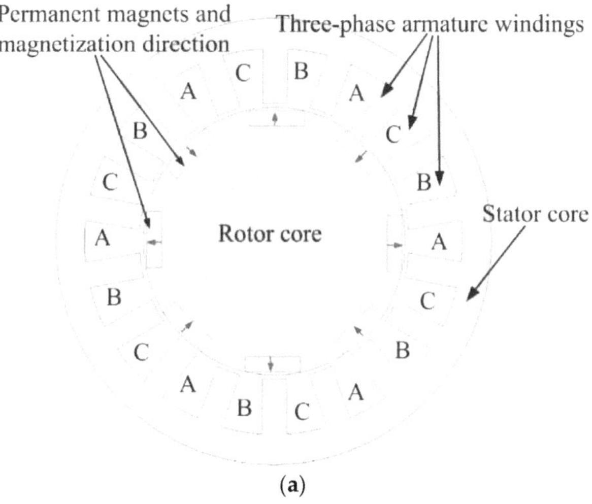


(a)

**Figure 9.** *Cont.*

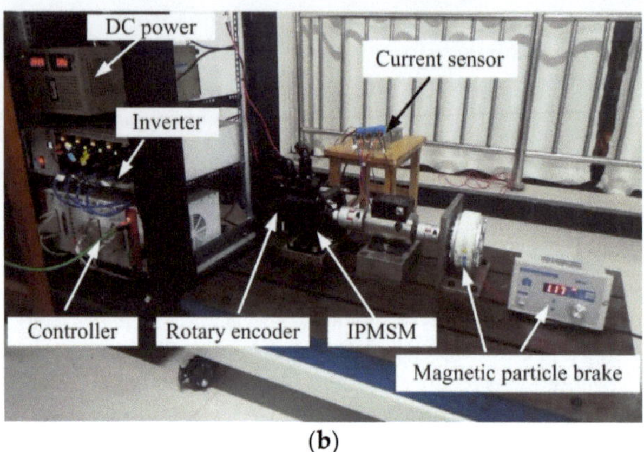



(**b**)

**Figure 9.** Experimental test of the dq-axis current reaction time of IPMSM: (**a**) IPMSM's structure; (**b**) test platform.

In order to keep the same operating conditions as the simulation model of Section 4.1, a magnetic particle brake is installed in the test platform of Figure 9b, and the magnetic particle brake is coaxial with the front-end department of the rotor.

During the experimental process, the experimental data were measured by a current sensor and processed by the software of the controller. If the three-phase currents of IPMSM were measured by an oscilloscope, the measured data looks more like the experimental data, but some calculation processes of Park transformation (from three-phase currents to dq-axis currents) should be achieved. Therefore, this paper adopts the current sensor and controller's software to measure and process the experimental data.

Figures 10 and 11 show the hardware test results of the dq-axis current reaction time of IPMSM by the PI method and IMC-PI method, where the inverter's switching cycle $T_{pwm}$ are 0.001 s and 0.1 s, respectively. With the increase in the inverter's switching cycle $T_{pwm}$, Figures 10a and 11a indicate that the response time of the q-axis current of IPMSM is 25 ms ($T_{pwm} = 0.001s$) and 50 ms ($T_{pwm} = 0.01s$), respectively. However, if the IMC-PI method is adopted, the response time of the q-axis current of IPMSM is nearly stable regardless of whether the inverter's switching cycle is $T_{pwm} = 0.001s$ or $T_{pwm} = 0.01s$, which are shown in Figures 10b and 11b.

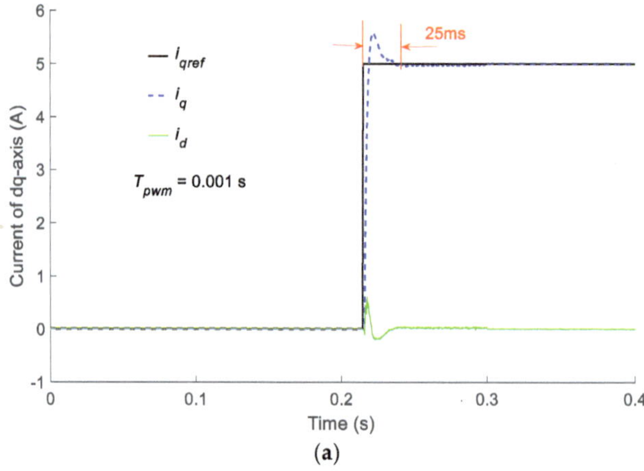



(**a**)

**Figure 10.** *Cont.*

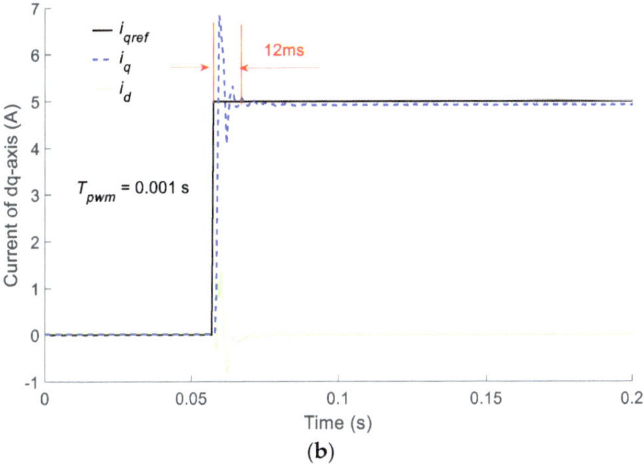



**Figure 10.** The hardware test results of the dq-axis current reaction time of IPMSM ($T_{pwm} = 0.001s$): (**a**) PI method; (**b**) IMC-PI method.

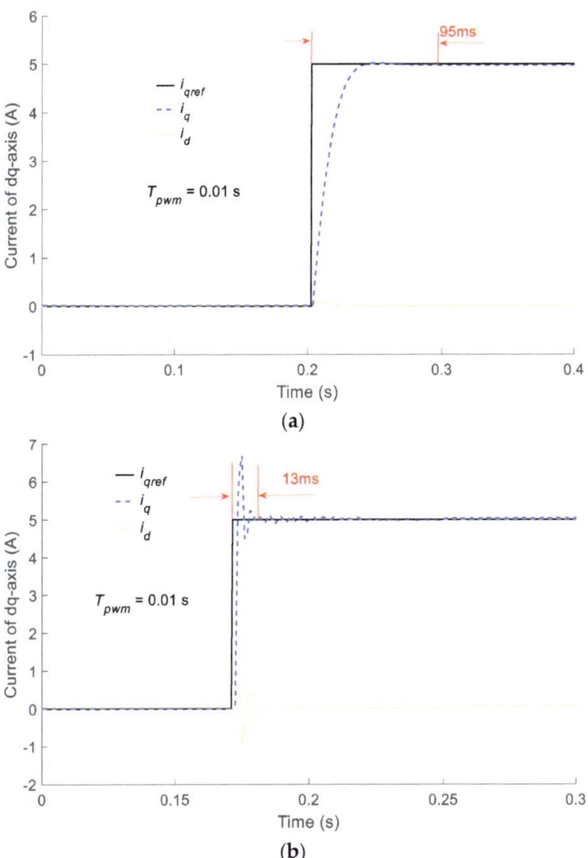



**Figure 11.** The hardware test results of the dq-axis current reaction time of IPMSM ($T_{pwm} = 0.01s$): (**a**) PI method; (**b**) IMC-PI method.

The comparison between Figures 6 and 10 shows that, in the condition of the inverter's switching cycle $T_{pwm} = 0.001s$, there is no significant difference in the hardware test result and simulation result. For example, by the PI method, the q-axis current reaction time of the hardware test and simulation are 25 ms and 28 ms, respectively. In another example, the q-axis current reaction time of the hardware test and simulation are 12 ms and 13 ms by adopting the IMC-PI method.

However, with the increasing of the inverter's switching cycle $T_{pwm}$, a difference in the q-axis current reaction time between the hardware test result and simulation result occurs. Take the inverter's switching cycle $T_{pwm} = 0.01s$ and PI method as an example, the q-axis current reaction time of the hardware test and simulation are 95 ms and 135 ms, respectively. Compared with the PI method, the IMC-PI method still has a good performance, in that the hardware test result (13 ms) agrees with the simulation result (13 ms).

Therefore, the above comparison of the hardware test result and simulation result vividly demonstrates that the dq-axis current reaction time of IPMSM is improved by the IMC-PI method.

Furthermore, because the larger inverter's switching cycle $T_{pwm}$ was not beneficial to the operation of the inverter (such as in the condition of $T_{pwm} = 0.1s$), the relevant hardware experimental test was not conducted in this paper.

## 5. Discussion

In Sections 4.1 and 4.2, the simulation results (see Figures 6 and 7) are compared and verified by hardware experiments (see Figures 10 and 11), and the hardware experiment results indicate that, based on the IMC-PI method, the stability time of the q-axis current start-up response of IPMSM was approximately 12 ms~13 ms, which was better than that of the PI method (25 ms~95 ms). However, the limitations of the proposed IMC-PI method and PI method should be discussed.

Firstly, the dq-axis current start-up response of IPMSM is only an intermediate quantity, and the final dq-axis control effect will be reflected in the speed and torque output of IPMSM.

Secondly, it is not that the shorter the stability time of dq-axis current start-up response of IPMSM the better, because the high overshoot of dq-axis current will cause the severe speed and torque output fluctuation of IPMSM, which is not beneficial to the safe operation and speed change of electric vehicles.

Thirdly, the results of the simulation and hardware experiment also demonstrate that the inverter's switching frequency has a significant influence on the parameter variation of the PI controller, thus affecting the control performance of the dq-axis current of IPMSM.

Therefore, in the next research work, inverter non-linearities (switching frequency), speed, and torque output of IPMSM should be fully considered, and the parameter variation of the PI method and IMC-PI method should be further researched.

## 6. Conclusions

Two methods were used in this paper to investigate the dq-axis current reaction time of IPMSM. After the theoretical model and controller design, the current-loop decoupling of IPMSM was presented, and the dq-axis current reaction time of IPMSM was simulated and analyzed. Then, the simulation results were verified by the hardware experiments, and the results of the simulation and hardware experiment indicated that the switching frequency of the inverter had a significant influence on the parameters setting of the PI controller. Lastly, the results of the simulation and hardware experiment demonstrated that, compared with the PI method, the dq-axis current reaction time of IPMSM was improved by the IMC-PI method. In addition, the results of the simulation and hardware experiment also showed that the IMC-PI method had the advantage of the easy setting of the control parameters and was not influenced by the inverter's switching frequency.

**Author Contributions:** Conceptualization, A.H. and Z.C.; methodology, A.H. and Z.C.; software, A.H. and J.W.; validation, Z.C.; writing—original draft preparation, Z.C.; writing—review and editing, A.H. All authors have read and agreed to the published version of the manuscript.

**Funding:** This work was financially supported by the Scientific and Technological Project in Henan Province under Grant No. 232102241024.

**Institutional Review Board Statement:** Not applicable.

**Informed Consent Statement:** Not applicable.

**Data Availability Statement:** Some or all data and models generated or used during the study are available in a repository or online.

**Conflicts of Interest:** The authors declare no conflict of interest.

## References

1. Uddin, M.N.; Rahman, M.M.; Patel, B.; Venkatesh, B. Performance of a loss model based nonlinear controller for IPMSM drive incor-porating parameter uncertainties. *IEEE Trans. Power Electron.* **2019**, *34*, 5684–5696. [CrossRef]
2. Shah, S.H.; Wang, X.; Abubakar, U.; Wang, L.; Gao, P. Investigation of noise and vibration characteristics of an IPMSM with modular-type winding arrangements having three-phase sub-modules for fault-tolerant applications. *IET Electr. Power Appl.* **2022**, *16*, 248–266. [CrossRef]
3. Alvaro-Mendoza, E.; Morales JD, L.; Hamida, M.A.; Ghanes, M. Angular position estimation error extraction for speed and angular position estimation of IPMSM using a parameter-free adaptive observer. *J. Frankl. Inst.* **2022**, *359*, 7140–7164. [CrossRef]
4. Zahraoui, Y.; Zaihidee, F.M.; Kermadi, M.; Mekhilef, S.; Mubin, M.; Tang, J.R.; Zaihidee, E.M. Fractional Order Sliding Mode Controller Based on Supervised Machine Learning Techniques for Speed Control of PMSM. *Mathematics* **2023**, *11*, 1457. [CrossRef]
5. Dong, Z.; Liu, C.; Liu, S.; Song, Z. Deadbeat Predictive Current Control for Series-Winding PMSM Drive with Half-Bridge Power Module-Based Inverter. *Energies* **2021**, *14*, 4620. [CrossRef]
6. Shuang, B.; Zhu, Z.Q. A novel method for estimating the high frequency incremental dq-axis and cross-coupling induct-ances in interior permanent magnet synchronous machines. *IEEE Trans. Ind. Appl.* **2021**, *57*, 4913–4923. [CrossRef]
7. Miguel-Espinar, C.; Heredero-Peris, D.; Gross, G.; Llonch-Masachs, M.; Montesinos-Miracle, D. Maximum torque per voltage flux-weakening strategy with speed limiter for PMSM drives. *IEEE Trans. Ind. Electron.* **2020**, *68*, 9254–9264. [CrossRef]
8. Guo, J.; Fan, T.; Li, Q.; Wen, X. Coupling and digital control delays affected stability analysis of permanent magnet syn-chronous motor current-loop control. In Proceedings of the IEEE Vehicle Power and Propulsion Conference, Hanoi, Vietnam, 14–17 October 2019.
9. Hernandez, O.S.; Cervantes-Rojas, J.S.; Oliver, J.P.O.; Castillo, C.C. Stator Fixed Deadbeat Predictive Torque and Flux Control of a PMSM Drive with Modulated Duty Cycle. *Energies* **2021**, *14*, 2769. [CrossRef]
10. Xu, L.; Chen, G.; Li, G.; Li, Q. Model Predictive Control Based on Parametric Disturbance Compensation. *Math. Probl. Eng.* **2020**, *2020*, 9543928. [CrossRef]
11. Son, D.-K.; Kwon, S.-H.; Kim, D.-O.; Song, H.-S.; Lee, G.-H. Control Comparison for the Coordinate Transformation of an Asymmetric Dual Three Phase Synchronous Motor in Healthy and Single-Phase Open Fault States. *Energies* **2021**, *14*, 1735. [CrossRef]
12. Yosuke, M.; Shinji, D. High Response Torque Control IPMSM Using State Feedback Control Based N-T Coord. *System. Electr. Eng. Jpn.* **2019**, *206*, 51–62.
13. Zhou, S.; Liu, J.; Zhou, L.; Zhang, Y. DQ current control of voltage source converters with a decoupling method based on preprocessed reference current feed-forward. *IEEE Trans. Power Electron.* **2017**, *32*, 8904–8921. [CrossRef]
14. Ma, Y.; Yang, X.; Zhou, X.; Yang, L.; Zhou, Y. Dual Closed-Loop Linear Active Disturbance Rejection Control of Grid-Side Converter of Permanent Magnet Direct-Drive Wind Turbine. *Energies* **2020**, *13*, 1090. [CrossRef]
15. Qiu, T.; Wen, X.; Zhao, F.; Wang, Y. Permanent magnet flux linkage adaptive observer for permanent magnet synchronous motor. In Proceedings of the IEEE Transportation Electrification Asia-Pacific, Beijing, China, 31 August–3 September 2014.
16. Qiu, T.; We, X.; Zhao, F. Design strategy of permanent magnet flux linkage adaptive observer for permanent magnet synchronous motor. *Proc. Chin. Soc. Electr. Eng.* **2015**, *35*, 2287–2294.
17. Saleh, M.; Hamid, Y.; Mojtaba, A.K. Adaptive sliding-mode type-2 neuro-fuzzy control of an induction motor. *Expert Syst. Appl.* **2015**, *42*, 6635–6647.
18. Moradi, H.; Yaghobi, H.; Alinejad-Beromi, Y.; Bustan, D. Power-control and speed-control modes of a DFIG using adaptive sliding mode type-2 neuro-fuzzy for wind energy conversion system. *IET Renew. Power Gener.* **2020**, *14*, 2946–2954. [CrossRef]
19. Xu, W.; Junejo, A.K.; Liu, Y.; Hussien, M.G.; Zhu, J. An Efficient Antidisturbance Sliding-Mode Speed Control Method for PMSM Drive Systems. *IEEE Trans. Power Electron.* **2020**, *36*, 6879–6891. [CrossRef]
20. Ahmed, W.A.E.M.; Adel, M.M.; Taha, M.; Saleh, A.A. PSO technique applied to sensorless field-oriented control PMSM drive with discretized RL-fractional integral. *Alex. Eng. J.* **2021**, *60*, 4029–4040. [CrossRef]

21. Eshwar, K.; Kumar, T.V. Reduction of torque and flux ripples in direct torque control for three-level open-end winding PMSM drive. *IET Electr. Power Appl.* **2020**, *14*, 2843–2854. [CrossRef]
22. Heidari, R. Model predictive combined vector and direct torque control of SM-PMSM with MTPA and constant stator flux magnitude analysis in the stator flux reference frame. *IET Electr. Power Appl.* **2020**, *14*, 2283–2292. [CrossRef]
23. Sreejith, R.; Singh, B. Sensorless Predictive Current Control of PMSM EV Drive Using DSOGI-FLL Based Sliding Mode Observer. *IEEE Trans. Ind. Electron.* **2020**, *68*, 5537–5547. [CrossRef]
24. Bensalem, Y.; Kouzou, A.; Abbassi, R.; Jerbi, H.; Kennel, R.; Abdelrahem, M. Sliding-Mode-Based Current and Speed Sensors Fault Diagnosis for Five-Phase PMSM. *Energies* **2021**, *15*, 71. [CrossRef]
25. Zhu, S.; Huang, W.; Lin, X.; Jiang, W.; Dong, D.; Wu, X. Simplified model predictive torque control for permanent magnet synchronous motor in static coordinate system. In Proceedings of the IEEE International Symposium on Predictive Control of Electrical Drives and Power Electronics, Quanzhou, China, 31 May–2 June 2019.
26. Lemmens, J.; Vanassche, P.; Driesen, J. PMSM Drive Current and Voltage Limiting as a Constraint Optimal Control Problem. *IEEE J. Emerg. Sel. Top. Power Electron.* **2014**, *3*, 326–338. [CrossRef]
27. Yuan, L.; Hu, B.X.; Wei, K.Y.; Chen, S. *The Control Principle and MATLAB Simulation of Modern Permanent Magnet Synchronous Motor*; Beijing University of Aeronautics and Astronautics Press: Beijing, China, 2016.
28. Ricardo, H. *A Modern Introduction to Differential Equations*; Elsevier Press: Amsterdam, The Netherlands, 2015.
29. Chen, B. *Electric Drive Automatic Control System: Motion Control System*, 3rd ed.; China Machine Press: Beijing, China, 2007.
30. Gu, C.; Chen, Z. *Electric Drive Automatic System and MATLAB Simulation*; Tsinghua University Press: Beijing, China, 2011.
31. Gene, F.F. *Principles and Design of Automatic Control*, 8th ed.; Publishing House of Electronics Industry: Beijing, China, 2021.
32. Hu, S. *Automatic Control Principle*, 6th ed.; Science Press: Beijing, China, 2015.

*World Electric Vehicle Journal*

*Article*

# Temperature Field Calculation of the Hybrid Heat Pipe Cooled Permanent Magnet Synchronous Motor for Electric Vehicles Based on Equivalent Thermal Network Method

Huimin Wang [1], Chujie Zhang [1], Liyan Guo [2], Wei Chen [1,*] and Zhen Zhang [2]

[1] School of Electrical Engineering, Tiangong University, Tianjin 300387, China; wanghuimin@tiangong.edu.cn (H.W.); 2031060797@tiangong.edu.cn (C.Z.)
[2] Advanced Electrical Equipment Innovation Center, Zhejiang University, Hangzhou 311107, China; 1012203006@tju.edu.cn (L.G.); zzhen@zju.edu.cn (Z.Z.)
* Correspondence: chen_wei@tju.edu.cn; Tel.: +86-022-83955692

**Abstract:** A hybrid heat-pipe cooling structure for the permanent magnet synchronous motor for electric vehicles was analyzed in this paper to effectively equalize the axial temperature rise and reduce the average temperature of each heat-generating component in the motor. A temperature field calculation method for the hybrid heat-pipe cooling permanent magnet synchronous motor based on an equivalent thermal network method was proposed in this paper to save computing resources in temperature field analyses for hybrid heat-pipe cooling motors. The results were verified against the simulation results of the computational fluid dynamics method under three common operating conditions for electric vehicles. The error was proven to be within 5%. The calculation time of the proposed method was compared with the computational fluid dynamics method, which demonstrated that the calculation time of the proposed method was within 194 s.

**Keywords:** permanent magnet synchronous motor; hybrid heat-pipe cooling structure; equivalent thermal network method; temperature field calculation

**Citation:** Wang, H.; Zhang, C.; Guo, L.; Chen, W.; Zhang, Z. Temperature Field Calculation of the Hybrid Heat Pipe Cooled Permanent Magnet Synchronous Motor for Electric Vehicles Based on Equivalent Thermal Network Method. *World Electr. Veh. J.* **2023**, *14*, 141. https://doi.org/10.3390/wevj14060141

Academic Editor: Zhongxian Chen

Received: 28 April 2023
Revised: 18 May 2023
Accepted: 23 May 2023
Published: 27 May 2023



## 1. Introduction

With the development of electric vehicle technology, its advantages, such as low energy consumption and no harmful emissions. have gradually made it a major transportation alternative to traditional internal combustion engine vehicles, providing an effective solution to the problem of human carbon neutrality [1]. Currently, common electric vehicle drive motors include asynchronous motors and permanent magnet synchronous motors (PMSM). With their high torque density and high efficiency, the PMSM has become the main traction motor used in electric vehicles such as the Tesla Model 3 and Porsche Taycan [2]. With the gradual maturity of electric vehicle technology, the permanent magnet synchronous motor as a drive motor is also developing in the direction of higher torque density and higher efficiency, which leads to a sharp increase in the temperature rise of the internal heating components of the motor, and the design of a reasonable cooling structure for it has become a hot research topic today.

In order to realize a rational design of the motor cooling system, an accurate calculation technique of the motor temperature field is an important tool in motor design work. Common temperature field calculation methods for permanent magnet synchronous motors mainly use commercial finite volume method temperature field calculation procedures based on computational fluid dynamics (CFD) [3], a main problem of which is that the calculation of complex motor cooling structures is time-consuming and requires a lot of computational resources. Such problems are particularly evident in the calculation of the motor temperature field of phase-change cooling structures [4], and it has become particularly important to improve the efficiency of the calculation of the motor temperature field of phase-change cooling with the development of phase-change materials and high-performance thermal conductivity materials such as heat pipes.

The equivalent thermal network method is a highly efficient method to calculate the temperature field of a motor, which is based on the principle of calculating the thermal resistance in the internal heat-transfer path of the motor to generate a thermal resistance network [5,6]. It uses a nodal voltage method similar to the circuit, with the losses generated by the internal heat-generating parts of the motor as the input quantity and the external ambient temperature as the reference node; the temperature rise of each heat-generating part of the motor is finally calculated after several iterations. J. Nerg et al. focused on a T-shaped equivalent thermal resistance network [7]. The advantages and disadvantages of the conventional thermal resistance network and the T-shaped equivalent thermal network were compared in the paper. The equivalent thermal network model of a surface-mounted permanent magnet synchronous motor was built by Si J et al. [8]. The improved thermal network model can derive the motor's temperature field distribution faster without complete motor parameters [9–11]. Only axial and radial heat transfer is generally concerned in the temperature field analysis of permanent magnet synchronous motors for electric vehicles [12]. Dawei Liang et al. proposed a two-dimensional thermal resistance network model that considers both axial and radial heat transfer and analyzed its effectiveness in improving the efficiency of motor temperature field calculation with guaranteed accuracy [13]. The temperature field of a permanent magnet synchronous motor for an unmanned aircraft was analyzed by Zhang C et al. The fixed heat pipe cools the motor on winding. The temperature field of the heat-pipe cooled motor was calculated based on the equivalent thermal network method [14]. Liu L et al. analyzed and calculated the temperature field of a permanent magnet synchronous motor for electric vehicles with fixed heat-pipe cooling using the equivalent thermal network method. The paper considered three common operating conditions of electric vehicles and compared the calculation results with CFD simulation results [15]. Sun D et al. analyzed the phase-change process inside the heat pipe in a heat-pipe cooling structure using the commercial CFD software Fluent [16], which analyzed the phase-change process of the fluid inside the heat pipe and the fluid motion state. Shi X et al. analyzed the state of the working mass inside the heat pipe in a rotating heat-pipe cooling structure for a high-speed machine tool motor using Fluent [17]. Subsequently, the correctness of their simulation analysis was proved using equivalent experiments by their research team [18].

According to the current research status, scholars mainly focus on calculating the temperature field of heat-pipe cooled motors for electric vehicles with fixed or rotating heat-pipe cooling structures. There is less research on analyzing the temperature field of motors with hybrid heat-pipe cooling structures based on the equivalent heat network method. The hybrid heat-pipe cooling structures include annular fixed and rotating heat pipes, and the equivalent thermal conductivity of rotating heat pipes varies with the motor speed, which leads to errors in the results obtained from the analytical calculation of the equivalent thermal network. The phase-change process of the internal mass during the operation of the rotating heat pipe cannot be analyzed by the equivalent thermal network method [19]. In this paper, a method to calculate the temperature field of the hybrid heat-pipe cooled motor based on the equivalent heat network method was proposed. The temperature field calculation and analysis were carried out for a hybrid heat-pipe cooled permanent magnet synchronous motor for electric vehicles using both fixed and rotating heat-pipe cooling. In this paper, the equivalent thermal conductivity of the fixed heat pipe and the rotating heat pipe under different operating conditions of the electric vehicle were obtained through the designed equivalent experiments, and the accuracy of the CFD calculation results was verified. The validity of the proposed temperature field calculation model was verified by comparing the model with the CFD simulation results. Finally, the computational time consumed by the proposed equivalent thermal network method of motor temperature field calculation was compared to the conventional CFD method in this paper.

## 2. Hybrid Heat-Pipe Cooling PMSM for EVs

### 2.1. Motor Model

The hybrid heat-pipe cooling structure adopted in the PMSM for EVs. The main features of the three-dimensional structure are shown in Figure 1.

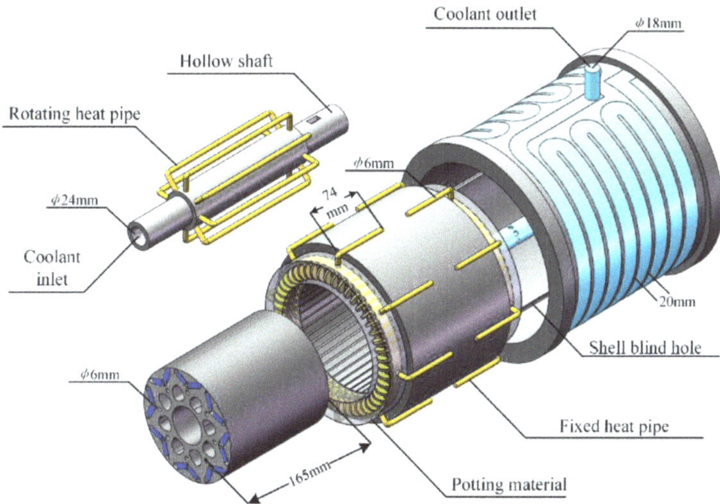


**Figure 1.** Schematic diagram of hybrid heat-pipe cooling structure.

It can be seen that there are two forms of heat pipes in the hybrid heat-pipe cooling structure, one is rotating and the other is fixed. An additional heat path from the rotor and permanent magnet to the water-cooled hollow shaft of the motor was built using the U-shaped rotating heat pipe installed inside the rotor, so that the large amount of heat generated by the rotor and permanent magnet when the motor is in high-speed operating conditions can be rapidly transferred to the cooling water inside the hollow shaft. The problem of the high average temperature rise of the rotor and permanent magnet, and uneven axial temperature rise, was solved effectively [20]. The additional heat path from the end winding to the motor water-cooled shell was constructed by using the ring-shaped fixed heat pipe, so that a large amount of heat generated by the winding can be quickly transferred to the motor-cooling water. The problem of the high temperature rise at the end of the winding was solved [21]. To further enhance the heat-transfer efficiency from the motor winding to the annular fixed heat pipe, a high thermal conductivity potting material was used to pot the winding end to the annular fixed heat pipe.

### 2.2. The Equivalent Thermal Conductivity of Hybrid Heat Pipe

The parameters of the hybrid heat-pipe cooling PMSM researched in this paper are shown in Table 1. In order to obtain the different thermal conductivity of the hybrid heat pipe under different working conditions of the hybrid heat-pipe cooling PMSM, the thermal conductivity and the internal mass motion state of the fixed and rotating heat pipe under different operating conditions of the PMSM were simulated by the CFD method. The VOF two-phase flow model and the LEE phase-change model were used to simulate the phase-change process and the internal mass motion of the rotating heat pipe at different speeds of the motor. The main working medium used in the working process of the rotating heat pipe involved in this paper was distilled water, the gaseous working medium was water vapor, and the phase-transition temperature of the liquid working medium inside the heat pipe was 15 °C.

**Table 1.** Parameter of hybrid heat-pipe cooling motor.

| Parameter | Date | Parameter | Date |
|---|---|---|---|
| Rated power | 90 kW | Pole-Slot number | 8p48s |
| Rated torque | 215 Nm | Pitch | 5 |
| Rated speed | 4000 r/min | Stator yoke diameter | 144 mm |
| Peak Power | 160 kW | Rotor yoke diameter | 48 mm |
| Maximum torque | 380 Nm | Air gap length | 0.5 mm |
| Maximum speed | 12,000 r/min | Polar arc coefficient | 0.64 |
| DC Bus Voltage | 320 V | Insulation level | H |

The volume fraction of the internal mass of the rotating heat pipe under rated working condition is shown in Figure 2. It can be seen that the liquid mass is mainly distributed in the radial top of the rotating heat pipe under the centrifugal force of rotor rotation.

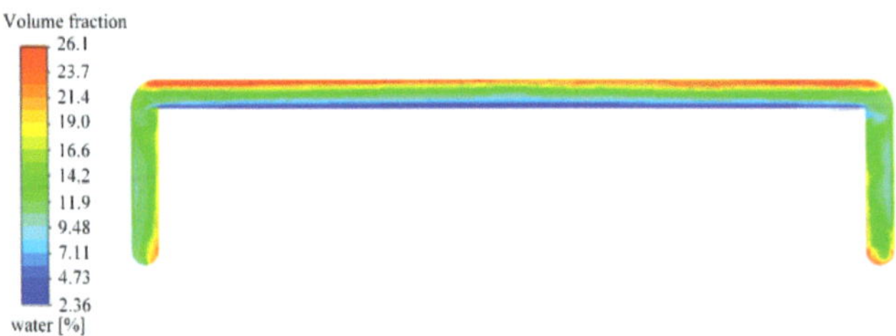


**Figure 2.** Cloud diagram of the volume fraction of the working mass inside the rotating heat pipe.

The temperature distribution cloud diagram of the rotating heat pipe at the rated working condition of the motor is shown in Figure 3. As shown in the diagram, the evaporation section of the heat pipe absorbs a large amount of heat generated by the rotor and permanent magnet, and the condensation section of the heat pipe effectively transfers the heat to the cooling water inside the hollow shaft of the motor.

**Figure 3.** Cloud diagram of temperature distribution of rotating heat pipe.

The equivalent thermal conductivity of the heat pipe calculated by the CFD method is shown in Table 2. It can be seen that when the PMSM for EVs is in the maximum torque working condition, that is, the speed is 600 r/min and the output torque is 380 N·m, the calculated thermal-resistance value is low. When the motor is at rated speed, due to the high motor speed, the fluid mass inside the rotating heat pipe is further accelerated back to the evaporation section by the centrifugal force of the rotor rotation, which further increases the equivalent thermal conductivity.

**Table 2.** Heat-pipe equivalent thermal conductivity simulation data.

| Motor Speed (rpm) | Thermal Resistance (°C/W) | Thermal Conductivity (W/m·k) |
| --- | --- | --- |
| 0 | 0.018 | 23,560.3 |
| 600 | 0.023 | 18,723.8 |
| 4000 | 0.006 | 74,628.2 |

## 3. Thermal Network Analysis of Hybrid Heat-Pipe Cooling Motor

### 3.1. Thermal Network Model Construction

The specific process of calculating the temperature field of the hybrid heat-pipe cooling motor by using the equivalent thermal network method is shown in Figure 4. Before the thermal network construction of the motor, the components with little influence on the temperature field calculation in the motor, such as the junction box, are generally simplified. After simplifying the motor model, the equivalent thermal network was established by treating the heating parts in the motor as nodes and the heat-transfer path as thermal resistance, and the thermal resistance of each heat-transfer path in the motor was calculated according to different working conditions and different speeds of the motor. The thermal resistance mainly included contact and convective thermal resistance between different materials. Then, the core loss of the stator and rotor, the eddy current loss of the permanent magnet, the copper loss of the winding, and the mechanical friction loss under three typical working conditions of the motor were calculated using FEA software and an empirical formula. These losses were taken as input values into the established equivalent thermal network. The node voltage method in the analogy circuit used the temperature of the external coolant as the reference node to obtain the cooling matrix. Finally, the cooling matrix was calculated and iteratively corrected until the temperature of each node, satisfying the residual conditions, was obtained.

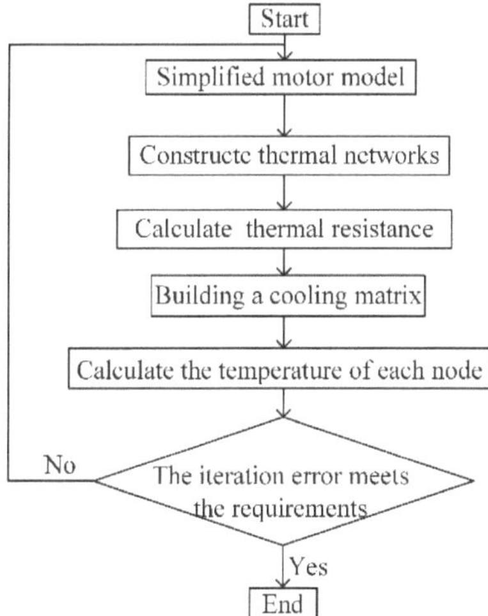


**Figure 4.** Thermal network establishment flowchart.

Firstly, the hybrid heat-pipe cooling motor is divided into thermal network nodes according to the materials of each heating component, and different nodes are connected by thermal resistance according to heat conduction and heat convection. In order to improve the calculation efficiency, the following assumptions are proposed under the premise of ensuring calculation accuracy [22,23]:

1. The motor's cooling effect and temperature distribution are circumferentially symmetrical;

2.  All the heat generated in the motor is regarded as all derived from the coolant, and the coolant temperature is regarded as the external node;
3.  Due to the heat pipe's relatively large thermal conductivity, heat pipe's additional heat path is regarded as heat conduction only through the heat pipe.
4.  All the losses of the rotor and the permanent magnet are transmitted from the heat pipe to the coolant in the hollow shaft.

Based on the above settings, the motor can be divided into stator yoke, stator teeth, winding, end winding heat pipe, thermal conductive potting material, rotor boot, permanent magnet, rotor yoke, rotor heat pipe, hollow shaft, bearing, end cover and shell according to the material and structure. Considering that the thermal conductivity of stator core and rotor core is low due to the stacking of silicon steel sheets in the axial direction, the stator core, rotor core, and permanent magnet are divided into three parts along the axial direction. Finally, the hybrid heat-pipe cooling motor is simplified to the equivalent node distribution diagram shown in Figure 5.

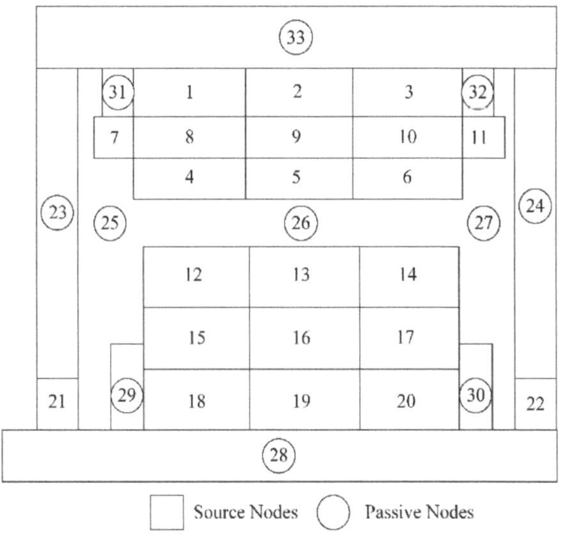


**Figure 5.** Node distribution in the motor.

According to the equivalent node diagram of the motor, the thermal resistance of heat conduction or thermal convection type can be added between the nodes, and the equivalent thermal network of the hybrid heat-pipe cooling permanent magnet synchronous motor for electric vehicle can be obtained. Among them, nodes 1–3 are stator yokes, nodes 7–11 are windings, nodes 4–6 are stator teeth, nodes 12–14 are rotor boots, nodes 18–20 are rotor yokes, nodes 15–17 are permanent magnets, nodes 21–22 are bearings, nodes 28 are hollow shafts, nodes 29–30 are rotor rotating heat pipes, nodes 25–27 are air domains, and nodes 23–24 are motor end covers. Node 31–32 is the heat pipe at the end winding, and node 33 is the water-cooled shell.

### 3.2. Thermal Resistance Calculation of Hybrid Heat-Pipe Cooling Structure

The heat transfer modes involved in the internal heating components of permanent magnet synchronous motor for electric vehicles include heat conduction, convective heat transfer and radiation heat transfer. This paper mainly analyzes the heat conduction and convective heat transfer inside the motor.

The form of heat conduction of permanent magnet synchronous motor for electric vehicles is mainly divided into plate heat conduction and cylinder heat conduction [24]. Most of the heat conduction in the motor is cylindrical heat conduction. The schematic diagram of the heat conduction principle on the cylinder wall is shown in Figure 6. For example, the stator yoke, the rotor yoke, the annular heat pipe and the rotating heat pipe can be equivalent to the heat conduction of the cylinder wall. The thermal resistance calculation method is as follows:

$$R = \frac{1}{2\pi l \lambda_1} \ln \frac{r_2}{r_1} + \frac{1}{2\pi l \lambda_2} \ln \frac{r_3}{r_2} \tag{1}$$

where, $l$ is the equivalent length of the cylinder parts such as the stator and rotor and the annular heat pipe, $\lambda_1$ is the thermal conductivity of the contact part 1, $\lambda_2$ is the thermal conductivity of the contact part 2, $r_1$ is the inner meridian of the cylinder, $r_2$ is the interface radius of the two-cylinder parts, and $r_3$ is the outer diameter of the cylinder.

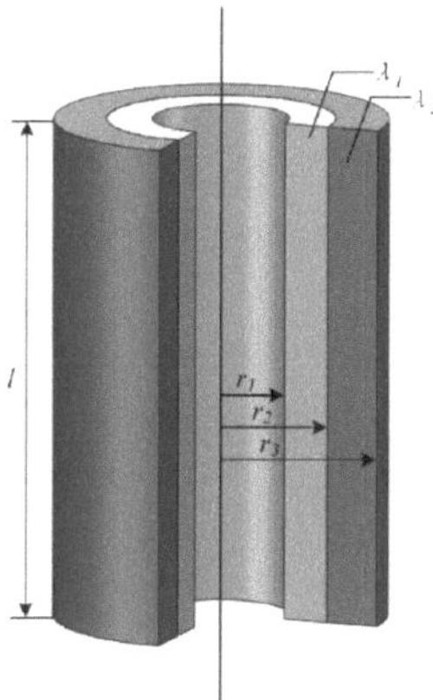

**Figure 6.** Schematic diagram of heat-conduction principle on the cylinder wall.

The air and cooling water in the motor belong to the fluid, and the fluid is mainly divided into laminar flow and turbulent flow according to its flow state [25,26]. In the motor, the critical Reynolds number is mainly used to judge the flow state of the fluid. It is generally believed that the fluid is laminar when the Reynolds number is less than 2200, and the fluid is turbulent when the Reynolds number is greater than 2200. The air in the air gap of the motor studied in this paper is in a turbulent state under the rated and maximum speed conditions, and the fluid in the air gap is in a laminar state under the maximum torque condition. As shown in Formula (7), the equivalent Reynolds number of the fluid in the motor can be calculated:

$$Re = u\frac{D}{\nu} \tag{2}$$

where, $D$ is the equivalent diameter of the cylinder, $v$ is the kinematic viscosity coefficient of the fluid, and $u$ is the velocity of the fluid.

According to the calculated equivalent Reynolds number, the Nusselt number of the fluid can be further calculated, and the convective heat transfer coefficient between the fluid and the solid can be finally obtained [27].

$$Nu = \begin{cases} 1.86 Re^{1/3} \mathrm{Pr}^{1/3} \left(\frac{D}{l}\right)^{1/3} \left(\frac{u}{v}\right)^{0.14} & Re < 2200 \\ 0.023 Re^{0.8} \mathrm{Pr}^{0.4} & Re > 2200 \end{cases} \tag{3}$$

$$\alpha = Nu\frac{\lambda}{D} \tag{4}$$

where, $\lambda$ is the thermal conductivity of the fluid, $Nu$ is the Nusselt number of the fluid, $Re$ is the Reynolds number of the fluid, $Pr$ is the Prandtl constant, $D$ is the equivalent diameter of the cylinder, $L$ is the equivalent length of the cylinder, $\alpha$ is the convective heat transfer coefficient.

According to the convection heat transfer coefficient of the fluid at different solid boundaries calculated above, the corresponding thermal convection thermal resistance can be obtained by substituting it into the following formula.

$$R = \frac{1}{S\alpha} \tag{5}$$

where, $S$ is the heat transfer area of the fluid-solid boundary.

The cooling structure of the hybrid heat pipe mainly includes the contact thermal resistance from the end winding to the annular fixed heat pipe, the convective thermal resistance from the rotor yoke to the cooling water in the hollow shaft through the U-shaped rotating heat pipe, and the equivalent thermal resistance of the annular heat pipe and the U-shaped rotating heat pipe itself. The three types of thermal resistance in the hybrid heat-pipe cooling structure are calculated below:

1. The thermal resistance from the end winding to the annular fixed heat pipe is $R_{87}$ and $R_{1011}$. Taking $R_{87}$ as an example:

$$R_{87} = \frac{d_{\mathrm{cudb}}}{2\lambda_{\mathrm{cu}} S_{87}} + \frac{1}{\lambda_{\mathrm{hp}} S_{87}} \tag{6}$$

where, $d_{cudb}$ is the radial thickness of the end winding, and $\lambda_{hp}$ is the static thermal conductivity of the heat pipe.

2. The thermal resistance of the rotor yoke through the rotating heat pipe to the cooling water inside the hollow shaft includes $R_{1828}$, $R_{1928}$ and $R_{2028}$. The thermal resistance $R_{1828}$ is analyzed as an example:

$$R_{1828} = \frac{1}{2\pi\lambda_{\mathrm{stj}} l/3} \ln\frac{r_{\mathrm{re}}}{r_{\mathrm{m}}} + \frac{1}{\alpha_{\mathrm{q}} S_{1828}} \tag{7}$$

where, $\alpha_{\mathrm{q}}$ is the convective heat transfer coefficient between the heat pipe and the cooling water, and $S_{1828}$ is the equivalent heat transfer area between the rotating heat pipe and the cooling water.

3. The thermal resistance of the annular fixed heat pipe at the end of the winding and the rotating heat pipe at the rotor yoke can be obtained by the temperature difference between the evaporation section and the condensation section of the heat pipe obtained by the CFD simulation above, namely:

$$R = \frac{T_{\mathrm{E}} - T_{\mathrm{c}}}{P} \tag{8}$$

where, $T_e$ is the steady-state temperature of the evaporation section of the heat pipe, $T_c$ is the steady-state temperature of the condensation section of the heat pipe, $P$ is the heat load of the evaporation section of the heat pipe, and $R$ is the equivalent thermal resistance of the heat pipe.

### 3.3. Thermal Resistance Calculation of Other Parts

#### 3.3.1. The Stator Yoke

According to the above thermal network division and node selection inside the motor, it can be seen that the stator yoke mainly contains five heat-transfer paths, which are the stator yoke to the winding, the stator yoke to the shell, the stator yoke to the potting glue, the stator yoke to the stator tooth, and the axial heat conduction of the stator yoke itself.

1. The thermal resistance from the stator yoke to the winding in the slot is $R_{18}$, $R_{29}$ and $R_{310}$. Considering the consistency of thermal resistance, $R_{18}$ is taken as the analysis object:

$$R_{18} = \frac{d_{\mathrm{de}}/2}{\lambda_{\mathrm{stj}} S_{18}} + \frac{d_{\mathrm{jy}}}{\lambda_{\mathrm{jy}} S_{18}} + \frac{d_{\mathrm{cu}}/2}{\lambda_{\mathrm{cu}} S_{18}} \tag{9}$$

where, $\lambda_{stj}$ is the radial thermal conductivity of the silicon–steel sheet in the stator yoke, $\lambda_{jy}$ is the thermal conductivity of insulation in the slot, $\lambda_{cu}$ is the thermal conductivity of the winding"s copper wire, $d_{de}$ is the radial thickness of the stator yoke, $d_{cu}$ is the radial thickness of the winding, $S_{18}$ is the equivalent heat-transfer area between node 1 and node 8, and the same below.

2. Taking the thermal resistance from the stator yoke to shell $R_{133}$ from node 1 to node 33 as an example:

$$R_{133} = \frac{1}{2\pi\lambda_{stj}\frac{l}{3}} \ln \frac{r_{dw}}{r_{den}} + \frac{1}{2\pi\lambda_{sh}\frac{l}{3}} \ln \frac{r_{sh}}{r_{dw}} \tag{10}$$

where, $\lambda_{sh}$ is the shell thermal conductivity, $r_{den}$ is the inner diameter of the stator yoke, $r_{dw}$ is the outer diameter of the stator yoke, and $r_{sh}$ is the outer diameter of the motor shell.

3. The thermal resistance from the stator yoke to the potting material is $R_{131}$ and $R_{132}$, taking $R_{131}$ as an example:

$$R_{131} = \frac{1}{2}R_{12} + \frac{1}{\lambda_{gf}S_{131}} \tag{11}$$

where, $R_{12}$ is the axial thermal resistance of stator yoke, $\lambda_{gf}$ is the thermal conductivity of the potting adhesive, and $S_{131}$ is the equivalent heat-transfer area between the stator yoke and potting material.

4. The thermal resistances from the stator yoke to stator teeth are $R_{14}$, $R_{25}$ and $R_{36}$. Taking $R_{14}$ as an example:

$$R_{14} = \frac{(r_{dw} - r_{dn})/2}{\lambda_{stj}S_{14}} \tag{12}$$

where, $r_{dn}$ is the inner diameter of the stator.

5. The axial thermal resistance of the stator yoke is $R_{12}$ and $R_{23}$. Taking $R_{12}$ as an example:

$$R_{12} = \frac{l/3}{\lambda_{stz}S_{12}} \tag{13}$$

where, $\lambda_{stz}$ is the axial thermal conductivity of the stator silicon–steel sheet, and $l$ is the axial length of the stator.

### 3.3.2. The Winding

According to the above division of the motor thermal network and the selection of nodes, it can be seen that there are three main heat-transfer paths of the winding, namely: winding to stator yoke, winding to stator teeth, winding axial thermal resistance.

1. The thermal resistance from the winding to the stator yoke is, taking $R_{81}$ between node 8 and node 1 as an example:

$$R_{81} = R_{18} \tag{14}$$

2. The thermal resistance from the winding to the stator teeth includes $R_{84}$, $R_{95}$, and $R_{106}$, taking $R_{84}$ as an example:

$$R_{84} = \frac{d_{cu}/2}{\lambda_{cu}S_{84}} + \frac{d_{jy}}{\lambda_{jy}S_{84}} + \frac{d_{cu}/2}{\lambda_{stj}S_{84}} \tag{15}$$

where, $d_{dc}$ is the radial thickness of stator teeth.

3. The axial thermal resistance of the winding includes $R_{89}$ and $R_{910}$, taking $R_{89}$ as an example:

$$R_{89} = \frac{l/3}{\lambda_{cu}S_{89}} \tag{16}$$

### 3.3.3. The Stator Teeth

According to the above division of the thermal network inside the motor and the selection of nodes, it can be seen that the main heat-transfer paths of the stator teeth are divided into four: thermal resistance from the stator teeth to the stator yoke, thermal resistance from the stator teeth to the winding, axial thermal resistance from the stator teeth and the thermal resistance from the air-gap to the stator teeth.

1.  The thermal resistance from the stator teeth to the stator yoke includes $R_{41}$, $R_{51}$, and $R_{63}$. Taking $R_{41}$ as an example for calculation:

$$R_{41} = R_{14} \tag{17}$$

2.  The thermal resistance from the stator teeth to the winding includes $R_{48}$, $R_{59}$, and $R_{610}$, taking $R_{48}$ as an example:

$$R_{48} = R_{84} \tag{18}$$

3.  The axial thermal resistance of stator teeth includes $R_{45}$ and $R_{56}$. Taking $R_{45}$ as an example:

$$R_{45} = \frac{l/3}{\lambda_{\text{stz}} S_{45}} \tag{19}$$

4.  The thermal resistance from stator teeth to air gap includes $R_{426}$, $R_{526}$ and $R_{626}$. Taking $R_{426}$ as an example:

$$R_{426} = \frac{h_{\text{st}}}{2\lambda_{\text{stj}} S_{426}} + \frac{1}{\alpha_{\text{cq}} S_{426}} \tag{20}$$

where, $\alpha_{\text{cq}}$ is the convective heat transfer coefficient between the stator teeth and the air-gap.

### 3.3.4. The Rotor Boot

There are four primary heat-transfer paths of the rotor boot: rotor boot to air-gap, rotor boot to permanent magnet, rotor boot to rotor yoke, and axial heat conduction of the rotor boot.

1.  The thermal resistance from the rotor boot to the air gap includes $R_{1226}$, $R_{1326}$ and $R_{1426}$. Taking $R_{1226}$ as an example:

$$R_{1226} = \frac{1}{2\pi\lambda_{\text{stj}} l/3} \ln \frac{r_{\text{rw}}}{r_{\text{re}}} + \frac{1}{\alpha_{\text{rq}} S_{1226}} \tag{21}$$

where, $r_{\text{rw}}$ is the outer diameter of the rotor, $r_{\text{re}}$ is the outer diameter of the rotor yoke, and $\alpha_{\text{cq}}$ is the convective heat-transfer coefficient between the rotor boot and the air gap.

2.  The thermal resistance from rotor boot to the permanent magnet includes $R_{1215}$, $R_{1316}$ and $R_{1417}$. The thermal resistance $R_{1215}$ is analyzed as an example:

$$R_{1215} = \frac{d_{\text{pm}}/2}{\lambda_{\text{pm}} S_{1215}} + \frac{1}{2\pi\lambda_{\text{stj}} l/3} \ln \frac{r_{\text{rw}}}{r_{\text{re}}} \tag{22}$$

where, $\lambda_{\text{pm}}$ is the thermal conductivity of the permanent magnet and $d_{\text{pm}}$ is the thickness of the permanent magnet.

3.  The thermal resistance from the rotor boot to the rotor yoke includes $R_{1218}$, $R_{1319}$ and $R_{1420}$. Taking $R_{1218}$ as an example for analysis:

$$R_{1218} = \frac{1}{2\pi \frac{180}{360} \lambda_{\text{stj}} l/3} \ln \frac{r_{\text{rw}}}{r_{\text{m}}} \tag{23}$$

where, $r_{\text{m}}$ is the inner diameter of the rotor.

4.    The axial thermal resistance inside the rotor boot includes $R_{1213}$ and $R_{1314}$. Taking $R_{1213}$ as an example for analysis:

$$R_{1213} = \frac{l/3}{\lambda_{stz}S_{1213}}$$

(24)

### 3.3.5. The Permanent Magnet

There are three main heat-transfer paths of the permanent magnet in the motor: the permanent magnet to the rotor boot, the permanent magnet to rotor yoke and the axial heat conduction of the permanent magnet.

1.    The thermal resistance from permanent magnet to rotor boot includes $R_{1512}$, $R_{1613}$ and $R_{1714}$. Taking $R_{1512}$ as an example for analysis:

$$R_{1512} = R_{1215}$$

(25)

2.    The thermal resistance from the permanent magnet to the rotor yoke includes $R_{1518}$, $R_{1619}$ and $R_{1720}$. The thermal resistance $R_{1518}$ is analyzed as an example:

$$R_{1518} = \frac{d_{pm}/2}{\lambda_{pm}S_{1518}} + \frac{1}{2\pi\lambda_{stj}l/3}\ln\frac{r_{rew}}{r_m}$$

(26)

3.    The axial thermal resistance of permanent magnet includes $R_{1516}$ and $R_{1617}$. Taking $R_{1516}$ as an example for analysis:

$$R_{1516} = \frac{l/3}{\lambda_{pm}S_{1516}}$$

(27)

### 3.3.6. The Rotor Yoke

The primary heat-transfer path of the rotor yoke in the motor is divided into three: the rotor yoke to the rotor boot, the rotor yoke to the permanent magnet, and the rotor yoke axial heat conduction.

1.    The thermal resistance from the rotor yoke to the rotor boot includes $R_{1812}$, $R_{1913}$ and $R_{2014}$. The thermal resistance $R_{1812}$ is analyzed as an example:

$$R_{1812} = R_{1218}$$

(28)

2.    The thermal resistance from the rotor yoke to the permanent magnet includes $R_{1815}$, $R_{1916}$ and $R_{2017}$. The thermal resistance $R_{1815}$ is analyzed as an example:

$$R_{1815} = R_{1518}$$

(29)

3.    The axial thermal resistance of the rotor yoke includes $R_{1819}$ and $R_{1920}$. Taking $R_{1819}$ as an example for analysis:

$$R_{1819} = \frac{l/3}{\lambda_{stz}S_{1819}}$$

(30)

### 3.4. Solution of Equivalent Thermal Network Model

According to the principle introduced above, the heat conduction thermal resistance and heat convection thermal resistance between each node in the motor were calculated, respectively. Taking the loss of each active node as input, the outflow energy and inflow energy of each node in the motor are equal due to the conservation of energy. Therefore, the thermal balance equation for any node can be obtained by analogy with the node voltage method in the circuit.

$$-G(i,1)T(1) - \ldots + G(i,i)T(i) - \ldots G(i,n)T(n) = P(i)$$

(31)

where, G $(i, i)$ is the self-thermal conductivity of node $i$, and the remaining thermal conductivity is the mutual thermal conductivity.

Each node's heat-balance equations in the thermal network are listed to form a matrix.

$$
\begin{bmatrix}
G(1,1) \cdots -G(1,i) \cdots -G(1,n) \\
\vdots \\
-G(i,1) \cdots G(i,i) \cdots -G(i,n) \\
\vdots \\
-G(n,1) \cdots -G(n,i) \cdots G(n,n)
\end{bmatrix}
\begin{bmatrix}
T(1) \\
\vdots \\
T(i) \\
\vdots \\
T(n)
\end{bmatrix}
=
\begin{bmatrix}
P(1) \\
\vdots \\
P(i) \\
\vdots \\
P(n)
\end{bmatrix}
\tag{32}
$$

that is:

$$
GT = P \tag{33}
$$

where, $G$ is the thermal conductivity matrix, $T$ is the temperature rise matrix, and $P$ is the loss matrix.

The cooling matrix was modeled, and the loss value of each node was substituted into the iterative solution to obtain the temperature rise of each node.

## 4. The Temperature Field Analysis of Hybrid Heat-Pipe Cooling PMSM for EVs

### 4.1. Hybrid Heat-Pipe Cooling PMSM for Evs

Based on the equivalent thermal network proposed in this paper, the temperature field calculation of a hybrid heat-pipe cooling PMSM for Evs was analyzed. The main parameters of the PMSM are shown in Table 1. The rated speed is 4000 rpm, the maximum speed is 12,000 rpm, and the maximum output torque is 380 Nm. The material properties of the potting glue used to enhance the heat-transfer efficiency of the winding end are shown in Table 3. The material properties of the main modeling areas involved in the motor are shown in Table 4.

**Table 3.** Potting material properties.

| Property | Unit | Date |
|---|---|---|
| Thermal conductivity | W/m·k | 3.2–3.4 |
| Breakdown voltage | kV/mm | $\geq$15 |
| Working temperature | °C | −55–180 |

**Table 4.** Modeling area and material properties.

| Modeling Area | Material | Thermal Conductivity (W/m·k) |
|---|---|---|
| Shell | Iron | 39.2 |
| Stator yoke (radial/axial) | Silicon steel sheet | 38.7/3.7 |
| Rotor (radial/axial) | Silicon steel sheet | 38.7/3.7 |
| Winding | Copper | 385 |
| Slot insulation | Insulating paper | 0.2 |
| Permanent magnet | Nd-Fe-B | 9 |
| Shaft | Steel | 42 |
| Bearing | Steel | 42 |
| Cavity | Air | 0.026 |

This article analyzed the losses of various heating components in the motor before calculating the temperature field using an equivalent thermal network. The various losses in the motor mainly included copper loss, iron loss, permanent magnet eddy current loss, and mechanical loss. Due to the fact that the iron loss of the motor was mainly related to the harmonic magnetic field during actual operation, and, in order to make the subsequent calculation results of the motor's equivalent thermal network more accurate, this article used non-sinusoidal excitation collected from the actual inverter for loss analysis.

The corresponding losses of each node in the equivalent thermal network under non-sinusoidal excitation under three operating conditions are shown in Table 5. As is shown in Table 5, the losses of the motor under three typical working conditions were obtained by electromagnetic FEA and an empirical formula, respectively.

**Table 5.** The loss value of each node of the motor under non-sinusoidal excitation.

| Node | Loss at Rated Condition (W) | Loss at Maximum Torque Condition (W) | Loss at Maximum Speed Condition (W) |
|---|---|---|---|
| 1–3 | 254.26 | 23.1 | 2147.53 |
| 7–11 | 6749.9 | 15,835.8 | 6145.6 |
| 4–6 | 484.57 | 123 | 766.97 |
| 12–14 | 32.4 | 12.5 | 126.75 |
| 18–20 | 52.4 | 11.1 | 200.7 |
| 15–17 | 11.1 | 5.8 | 29.3 |
| 21–22 | 126.1 | 18.9 | 432 |

*4.2. Equivalent Thermal Conductivity Experiment of Heat Pipe*

The equivalent thermal conductivity of the heat pipe is obtained by designing a reasonable equivalent experiment to calculate the contact thermal resistance further accurately. The experimental platform is shown in Figure 7, including drive motor, PT100 thermocouple, rotating hollow shaft, DC heating wire, etc. This paper mainly analyzes the equivalent thermal conductivity of the rotating heat pipe at different speeds of the motor under the same thermal load of 200 W. The main principle of the experiment is to fix the U-shaped rotating heat pipe on the rotating shaft. By measuring the temperature difference between the evaporation section and the condensation section of the rotating heat pipe at different speeds, the equivalent thermal conductivity of the rotating heat pipe can be calculated by introducing the formula (34–35).

$$R = \frac{T_E - T_c}{p} \tag{34}$$

$$\Phi = \frac{\Delta t}{\frac{1}{Ak}} = \frac{\Delta t}{R} \tag{35}$$

where, $\Phi$ is the heat flux transmitted through the heat pipe, $T_e$ is the steady-state temperature of the evaporation section of the heat pipe, $T_c$ is the steady-state temperature of the condensation section of the heat pipe, $P$ is the heat load of the evaporation section of the heat pipe, $\Delta t$ represents the average temperature difference between the evaporation and condensation section, $R$ is the equivalent thermal resistance of the heat pipe, $k$ is the equivalent heat transfer coefficient of the heat pipe, and $A$ is the equivalent contact area of the heat pipe.

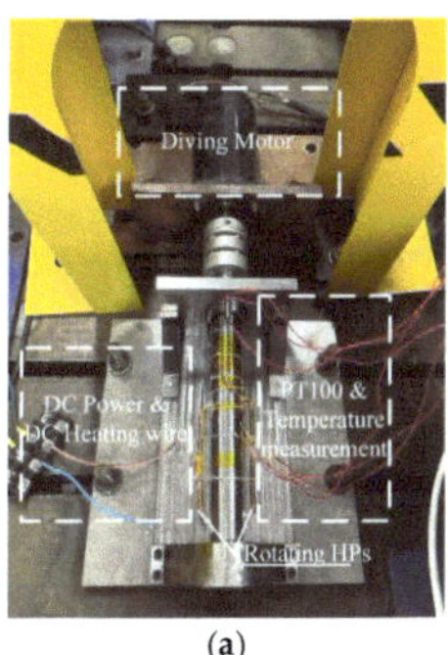


(a)

(b)

**Figure 7.** Equivalent thermal conductivity experimental equipment: (**a**) experimental platform; and (**b**) drive and measuring equipment.

The experimental data of the equivalent thermal conductivity of the heat pipe are shown in Table 6. When the heat load of the rotor is constant, the thermal conductivity of the rotating heat pipe increases first, then decreases, and finally increases again with the increase of the rotor speed. The highest thermal conductivity can reach 76,804.9 W/m·k at 4000 rpm.

**Table 6.** Experimental data of equivalent thermal conductivity of heat pipe.

| Motor Speed (rpm) | Temperature (°C) | | Thermal Resistance (°C/W) | Thermal Conductivity (W/m·k) |
|---|---|---|---|---|
| | $T_c$ | $T_e$ | | |
| 0 | 109.78 | 113.37 | 0.018 | 25,601.6 |
| 600 | 16.95 | 21.62 | 0.023 | 20,036.1 |
| 1200 | 22.71 | 25.12 | 0.012 | 38,402.5 |
| 1800 | 24.96 | 28.45 | 0.017 | 27,107.6 |
| 2400 | 26.40 | 30.31 | 0.019 | 24,254.2 |
| 3000 | 30.41 | 33.62 | 0.016 | 28,801.8 |
| 3600 | 32.59 | 34.96 | 0.012 | 38,402.5 |
| 4000 | 35.14 | 36.40 | 0.006 | 76,804.9 |

As shown in Table 7, the equivalent thermal conductivity of the heat pipe obtained by the CFD simulation iwa compared with the conclusion of the equivalent experimental calculation designed in this paper. It can be seen that the thermal conductivity of the rotating heat pipe obtained by the CFD simulation is low, which is due to the fact that CFD cannot effectively simulate the liquid working medium provided by the wick structure inside the heat pipe to further accelerate the reflux effect when simulating the rotating heat pipe.

**Table 7.** Comparison of thermal conductivity of rotating heat pipe.

| Speed (rpm) | Equivalent Experiments (W/m·k) | CFD Calculations (W/m·k) |
|---|---|---|
| 0 | 25,601.6 | 23,560.3 |
| 4000 | 76,804.9 | 74,628.2 |

*4.3. Equivalent Thermal Conductivity Experiment of Heat Pipe*

Aimed at the cooling matrix equation proposed above, the iterative calculation of the temperature rise was solved in each modeling area under three working conditions of the rated condition, maximum speed and maximum torque. The iterative calculation time of the program was between 105–194 s under the three working conditions. The temperature rise calculation results of each modeling area under non-sinusoidal excitation are shown in Table 8.

**Table 8.** Non-sinusoidal excitation temperature comparison.

| Operating Conditions | Calculation Method | Winding | Stator | PM | Rotor |
|---|---|---|---|---|---|
| | Thermal network | 89.34 °C | 44.32 °C | 35.12 °C | 33.98 °C |
| Rated condition | CFD | 86.48 °C | 45.4 °C | 35.54 °C | 34.33 °C |
| | Error | 3.2% | 2.4% | 1.2% | 1% |
| | Thermal network | 141.82 | 45.62 | 27.12 | 27.09 |
| Maximum torque condition | CFD | 138.77 | 46.55 | 27.8 | 27.62 |
| | Error | 2.2% | 2% | 2.4% | 2% |
| | Thermal network | 87.62 | 49.27 | 37.11 | 33.98 |
| Maximum speed condition | CFD | 85.26 | 49.81 | 37.72 | 34.59 |
| | Error | 2.7% | 1.1% | 1.6% | 1.8% |

Subsequently, the CFD method was used to calculate the temperature field of the hybrid heat-pipe cooling motor to verify the effectiveness of the proposed calculation method based on the equivalent thermal network method. When the motor is at rated and maximum speed conditions, the k-epsilon classical turbulence model and the standard wall function are used to compensate for the grid-quality degradation caused by the complex shape inside the motor and the double-layer full y + processing to compensate for the grid-quality degradation caused by the complex shape of the motor during the simulation of the turbulence model. For the maximum torque condition, the fluid state in the air-gap can be judged as a laminar flow state according to the Reynolds number calculated by the air-gap, so the classical laminar flow model and the standard wall function were used for the simulation calculation. The radiation heat process was ignored because of its lower inference in terms of the temperature field calculation.

Based on the loss analysis and internal fluid state analysis of permanent magnet synchronous motors for hybrid heat-pipe cooled electric vehicles mentioned above, the CFD method was used to verify the accuracy of the proposed equivalent heat network method calculation results. The comparison is shown in Table 8. The temperature distribution nephogram of each heating component in the motor under three working conditions under non-sinusoidal excitation is shown in Figure 8. According to Table 8, it can be seen that the maximum error is within 3.2%, and the error mainly occurs at the end winding. This is mainly because the influence of the low thermal conductivity of the potting material was ignored in modeling the equivalent effect thermal network. After analysis, it can be considered that the error of the equivalent thermal network model established in this article is within an acceptable range.

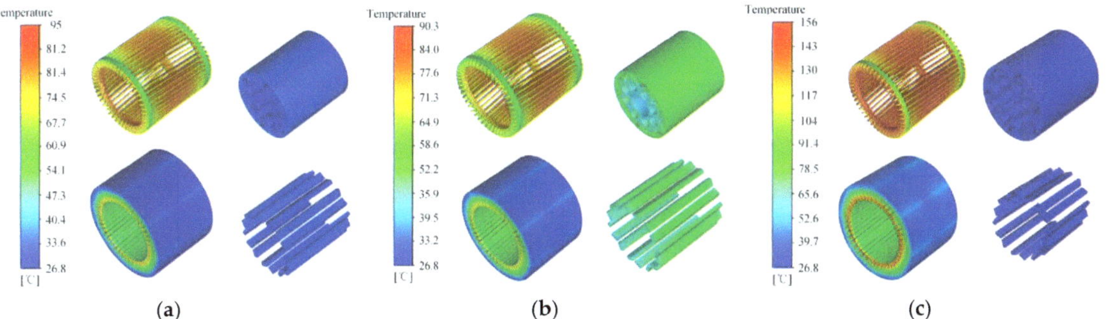

**Figure 8.** Temperature field under non-sinusoidal excitation: (**a**) rated condition; (**b**) maximum speed condition; and (**c**) maximum torque condition.

## 5. Conclusions

In this paper, the equivalent thermal network method was used to analyze and calculate the temperature field of the hybrid heat-pipe cooling permanent magnet synchronous motor for electric vehicle with a rated power of 90 kw. The equivalent thermal conductivity of the rotating heat-pipe structure was obtained by the equivalent experiment. The temperature calculation results of each node under three typical conditions of rated condition, maximum torque condition, and maximum speed condition were compared with the simulation analysis results, and the following conclusions were obtained:

1.  The equivalent thermal resistance of the rotating heat pipe decreases first, then increases, and finally decreases with the increase of the rotor rotation speed when the thermal load is constant. The thermal resistance is low at the rated speed of 4000 rpm, and the maximum equivalent thermal conductivity can reach 76,804.9W/m·k;

2.  By establishing an equivalent thermal network for the motor, the temperature of each node was analyzed and calculated. When comparing the calculation results with the CFD simulation results, it was found that the temperature error of each node was

within 5%, which is within the acceptable range. In view of the error, the analysis in this paper was mainly due to the complex shape of the cooling structure of the rotating heat pipe and the cooling structure of the heat pipe at the end winding, and the existence of potting glue between the end winding and the fixed heat pipe, which affects the accuracy of the thermal resistance calculation, so there is a certain error in the calculation results;

3.  In this paper, by comparing the temperature field calculation method of hybrid heat-pipe cooling motor based on equivalent thermal network with the traditional CFD simulation time, it was found that the iteration time of the proposed calculation method is within 194 s, which greatly shortens the temperature field calculation time of the hybrid heat-pipe cooling PMSM for EVs.

The equivalent thermal network method can be used to quickly calculate the temperature field distribution inside the hybrid heat-pipe cooling structure PMSM under the premise of ensuring a certain accuracy, which will be valuable as a reference for the related research of heat-pipe cooling structure motors in the future.

**Author Contributions:** Conceptualization, H.W.; methodology, H.W. and C.Z.; software, C.Z.; validation, C.Z.; formal analysis, C.Z.; investigation, L.G.; resources, H.W. and W.C.; data curation, L.G.; writing—original draft preparation, C.Z.; writing—review and editing, H.W. and Z.Z.; visualization, H.W. and Z.Z.; supervision, W.C.; project administration, W.C.; funding acquisition, H.W. and L.G. All authors have read and agreed to the published version of the manuscript.

**Funding:** This research was funded by "National Natural Science Foundation of China, grant number 52277064" and "Zhejiang Provincial Natural Science Foundation of China, grant number LY22E070005".

**Data Availability Statement:** The data presented in this study are available in the article.

**Conflicts of Interest:** There is no conflict of interest.

# References

1.  Wang, X.; Li, B.; Gerada, D.; Huang, K.; Stone, I.; Worrall, S.; Yan, Y. A critical review on thermal management technologies for motors in electric cars. *Appl. Therm. Eng.* **2022**, *117758*, 1359–4311. [CrossRef]
2.  Gronwald, P.O.; Kern, T.A. Traction Motor Cooling Systems: A Literature Review and Comparative Study. *IEEE Trans. Transp. Electrif.* **2021**, *7*, 2892–2913. [CrossRef]
3.  Dan, D.; Yao, C.; Zhang, Y.; Zhang, H.; Zeng, Z.; Xu, X. Dynamic thermal behavior of micro heat pipe array-air cooling battery thermal management system based on thermal network model. *Appl. Therm. Eng.* **2019**, *162*, 114183. [CrossRef]
4.  Vese, I.; Marignetti, F.; Radulescu, M.M. Multiphysics approach to numerical modeling of a permanent-magnet tubular linear motor. *IEEE Trans. Ind. Electron.* **2010**, *57*, 320–326. [CrossRef]
5.  Boglietti, A.; Cavagnino, A.; Staton, D.; Shanel, M.; Mueller, M.; Mejuto, C. Evolution and modern approaches for thermal analysis of electrical machines. *IEEE Trans. Ind. Electron.* **2009**, *56*, 871–882. [CrossRef]
6.  Tessarolo, A.; Bruzzese, C. Computationally efficient thermal analysis of a low-speed high-thrust linear electric actuator with a three-dimensional thermal network approach. *IEEE Trans. Ind. Electron.* **2015**, *62*, 1410–1420. [CrossRef]
7.  Nerg, J.; Rilla, M.; Pyrhonen, J. Thermal Analysis of Radial-Flux Electrical Machines with a High Power Density. *IEEE Trans. Ind. Electron.* **2008**, *55*, 3543–3554. [CrossRef]
8.  Si, J.; Zhao, S.; Feng, H.; Hu, Y.; Cao, W. Analysis of temperature field for a surface-mounted and interior permanent magnet synchronous motor adopting magnetic-thermal coupling method. *CES Trans. Electr. Mach. Syst.* **2018**, *2*, 166–174. [CrossRef]
9.  Li, L.; Zhang, J.; Zhang, C.; Yu, J. Research on Electromagnetic and Thermal Issue of High-Efficiency and High-Power-Density Outer-Rotor Motor. *IEEE Trans. Appl. Supercond.* **2016**, *26*, 5204805. [CrossRef]
10. Zhang, B.; Qu, R.; Xu, W.; Wang, J.; Chen, Y. Thermal model of totally enclosed water-cooled permanent magnet synchronous machines for electric vehicle applications. In Proceedings of the 2014 International Conference on Electrical Machines (ICEM), Berlin, Germany, 2–5 September 2014; pp. 2205–2211. [CrossRef]
11. Xiao, S.; Griffo, A. Online thermal parameter identification for permanent magnet synchronous machines. *Electr. Power Appl.* **2020**, *14*, 2340–2347. [CrossRef]
12. Huang, J.; Naini, S.S.; Miller, R.; Rizzo, D.; Sebeck, K.; Shurin, S.; Wagner, J. A Hybrid Electric Vehicle Motor Cooling System—Design, Model, and Control. *IEEE Trans. Veh. Technol.* **2019**, *68*, 4467–4478. [CrossRef]
13. Liang, D.; Zhu, Z.Q.; Zhang, Y.; Feng, J.; Guo, S.; Li, Y.; Wu, J.; Zhao, A. A hybrid lumped-parameter and two-dimensional analytical thermal model for electrical machines. *IEEE Trans. Ind. Appl.* **2020**, *57*, 246–258. [CrossRef]

14. Zhang, C.; Zhang, X.; Zhao, F.; Gerada, D.; Li, L. Improvements on Permanent Magnet Synchronous Motor by Integrating Heat Pipes into Windings for Solar Unmanned Aerial Vehicle. *Green Energy Intell. Transp.* **2022**, *1*, 100011. [CrossRef]
15. Liu, L.; Tan, G.; Zhou, F.; Sun, M.; Liu, Z.; Yang, C. A Study on Heat Dissipation of Electric Vehicle Motor Based on Heat-Pipe Heat Transfer Analysis. *SAE Technical Paper*, 6 April 2021.
16. Sun, D.; Xu, J.; Chen, Q. Modeling of the evaporation and condensation phase-change problems with FLUENT. *Numer. Heat Transf. Part B Fundam.* **2014**, *66*, 326–342. [CrossRef]
17. Sun, Y.; Zhang, S.; Chen, G.; Tang, Y.; Liang, F. Experimental and numerical investigation on a novel heat pipe based cooling strategy for permanent magnet synchronous motors. *Appl. Therm. Eng.* **2020**, *170*, 114970. [CrossRef]
18. Liang, F.; Gao, J.; Li, F.; Xu, L.; Wang, Z.; Jiang, H. A central cooling structure for motorized spindles: Principle and application. In Proceedings of the 2019 18th IEEE Intersociety Conference on Thermal and Thermomechanical Phenomena in Electronic Systems (ITherm), Las Vegas, NV, USA, 28–31 May 2019; pp. 1204–1211.
19. Chang, S.W.; Hsieh, M.F.; Wu, P.S.; Cai, W.L. Convective heat transfer motivated by liquid-to-vapor density difference in centrifugal force field of axially rotating loop thermosyphons. *Processes* **2021**, *9*, 1909. [CrossRef]
20. Wan, Z.; Sun, B.; Wang, X.; Wen, W.; Tang, Y. Improvement on the heat dissipation of permanent magnet synchronous motor using heat pipe. *Proc. Inst. Mech. Eng. Part D J. Automob. Eng.* **2020**, *234*, 1249–1259. [CrossRef]
21. Zhu, Z.Q.; Pang, Y.; Howe, D.; Iwasaki, S.; Deodhar, R.; Pride, A. Analysis of electromagnetic performance of flux-switching permanent-magnet machines by nonlinear adaptive lumped parameter magnetic circuit model. *IEEE Trans. Magn.* **2005**, *41*, 4277–4287. [CrossRef]
22. Mellor, P.H.; Roberts, D.; Turner, D.R. Lumped parameter thermal model for electrical machines of TEFC design. *IEE Proc. B (Electr. Power Appl.)* **1991**, *138*, 205–218. [CrossRef]
23. Demetriades, G.D.; de la Parra, H.Z.; Andersson, E.; Olsson, H. A real-time thermal model of a permanent-magnet synchronous motor. *IEEE Trans. Power Electron.* **2010**, *25*, 463–474. [CrossRef]
24. Park, J.B.; Moosavi, M.; Toliyat, H.A. Electromagnetic-thermal coupled analysis method for interior PMSM. In Proceedings of the 2015 IEEE International Electric Machines & Drives Conference (IEMDC), Coeur d'Alene, ID, USA, 10–13 May 2015; pp. 1209–1214.
25. Menter, F. Two-equation Eddy-viscosity Turbulence Models for Engineering Applications. *AIAA J.* **1994**, *32*, 1598–1605. [CrossRef]
26. Howey, D.A.; Childs PR, N.; Holmes, A.S. Air-gap convection in rotating electrical machines. *IEEE Trans. Ind. Electron.* **2010**, *59*, 1367–1375. [CrossRef]
27. Wang, H.; Liu, X.; Kang, M.; Guo, L.; Li, X. Oil injection cooling design for the IPMSM applied in electric vehicles. *IEEE Trans. Transp. Electrif.* **2022**, *8*, 3427–3440. [CrossRef]

MDPI AG
Grosspeteranlage 5
4052 Basel
Switzerland
Tel.: +41 61 683 77 34

*World Electric Vehicle Journal* Editorial Office
E-mail: wevj@mdpi.com
www.mdpi.com/journal/wevj